Yuri Rommel Vieira Araújo
Luiz Moreira Coelho Junior

Environmental assessment and forecasting of urban forestry waste

Yuri Rommel Vieira Araújo
Luiz Moreira Coelho Junior

Environmental assessment and forecasting of urban forestry waste

Urban forestry waste as an energy alternative

ScienciaScripts

Cover image: www.ingimage.com

This book is a translation from the original published under ISBN 978-613-9-63182-7.

Publisher:
Sciencia Scripts
is a trademark of
Dodo Books Indian Ocean Ltd. and OmniScriptum S.R.L publishing group

120 High Road, East Finchley, London, N2 9ED, United Kingdom
Str. Armeneasca 28/1, office 1, Chisinau MD-2012, Republic of Moldova, Europe
Printed at: see last page
ISBN: 978-620-7-79119-4

CONTENTS

I dedicate it to my parents, Jaciara and Aristóteles (*in memoriam*), who have always encouraged me and prioritized my studies.

ACKNOWLEDGMENTS

First of all, I thank God for life and for guiding me through this work. I would like to thank everyone who directly or indirectly contributed to the completion of this work:

To my parents, Jaciara and Aristóteles (*in memoriam*), and my stepfather Joao, for their life lessons, support and encouragement.

To my wife, friend and companion Zayne for her encouragement, patience and support in starting, continuing and finishing my master's degree, because without you I wouldn't have done it.

To Professor Luiz Moreira Coelho Junior for his guidance, support, dedication, patience and encouragement, without which this dissertation would not have been completed.

Professor Monica Carvalho for her guidance and knowledge acquired during the preparation of articles and classes.

To Zayne, Thiago Melquiades and Monijany for their help, assistance and preparation of the articles.

To my friends at the Directorate of Studies and Research (DIEP) and other friends at the Joao Pessoa Department of the Environment who supported, helped and encouraged me.

"Nobody hits you as hard as life. But it's not about how hard you can hit, it's about how much you can take and keep going. That's how victory is achieved."

Rock Balboa

"Satisfaction is in the effort, not in the achievement. Complete effort means complete victory.

We have to be the transformation we want in the world."

Mahatma Gandhi

"When we accept our limits, we can go beyond them.

A person who has never made a mistake, has never tried anything new."

Albert Einstein

SUMMARY

Urban tree planting generates a significant amount of solid biomass waste. This study carried out an environmental assessment and planned the use of waste from urban tree planting in Joao Pessoa as an energy alternative. More specifically, the objectives were: i) to diagnose and plan the maintenance of Joao Pessoa's urban tree planting; ii) to assess the environmental impacts and analyze different scenarios for Joao Pessoa's urban tree planting waste, with a view to inclusion in the Clean Development Mechanism (CDM), and iii) to analyze the forecast volume of waste from Joao Pessoa's urban tree planting. In order to achieve the objectives, an inventory of urban trees was carried out using simple random sampling, a Life Cycle Assessment (LCA) was developed using the IPCC 2013 GWP 100a method, and models from the ARIMA (*Autoregressive Integrated Moving Average*) family were applied. It was found that urban tree planting is irregularly distributed throughout the municipality, with the species *F. benjamina*, *S'. sıamea* and *T. catappa standing* out in terms of abundance and with the highest rates of phytopathological problems, physical damage and conflicts with urban facilities. The environmental assessment showed that the current disposal method used (landfill) has the greatest impact on the environment; the best scenario was to generate electricity, including the possibility of being used as carbon credits. The application of the ARIMA (0,1,4) model provided the best forecast for the 12 periods for the volume of urban tree waste in Joao Pessoa.

Keywords: biomass, forestry planning, life cycle assessment, energy use.

1. GENERAL INTRODUCTION

Urban afforestation can be defined as all the vegetation that makes up the urban landscape, and is divided into green areas (parks, woods, squares and gardens) and the afforestation of public roads (COPEL, 2009).

Urban afforestation provides many benefits for society. Through the environmental services it provides, it contributes to the formation of a healthy and comfortable physical environment, such as: regulating and purifying the air, soil and water; mitigating heat and stabilizing the climate; providing shelter and food for fauna; beautification; reducing noise pollution; circulating nutrients; protecting the soil from erosion; reducing levels of human stress; shade and leisure for the city's streets and avenues; among others (MELO; SEVERO, 2010; VALE et al., 2011).

For urban vegetation to remain in a good state of conservation, it is necessary to carry out periodic preventive and corrective actions (PIVETA; SILVA FILHO, 2002). These include irrigation during periods of low rainfall, actions to combat pests and diseases caused by insects and pathogens, and the maintenance and replacement of trees that present conflicts with public facilities or are a danger to the population (ARACRUZ, 2013). The aim of urban pruning is to manage the development of the upper part of trees (canopy), remove dry and diseased branches and generate solid waste (SÂO PAULO, 2015).

Solid urban waste includes household waste and urban cleaning waste (sweeping, weeding and urban forestry waste). In 2014, approximately 79 million tons of municipal solid waste (MSW) were generated in Brazil, an increase of 2.9% compared to the previous year (ABRELPE, 2015). In Paraiba, an average of 3,504 tons (t) of MSW were generated per day in 2014. Below Pernambuco, 8,830 t/day, and above Rio Grande do Norte, 3,009 t/day (ABRELPE, 2015).

In 2013, Joao Pessoa generated 263,000 tons of solid urban waste, considering only household and urban cleaning service waste (EMLUR, 2014a). In 2012, 9,200 trees were pruned or replaced in 64 neighborhoods and generated 26,243.80 t of biomass waste, resulting in high costs for maintaining the municipality's tree planting (JOÂO

PESSOA, 2012).

In cities with few financial resources and other emergency priorities, urban tree waste (biomass) can be an alternative source of income for many Brazilian municipalities (BARATTA JUNIOR; MAGALHÂES, 2010).

One of the tools available for verifying cost reductions in urban tree maintenance is the analysis of the periodic generation of biomass waste. By forecasting the generation of such waste, it is possible to verify its influence on operating costs, planning and the use of public resources.

The use of time series generates information about past behavior and is useful for the probable future. Using a model, past movements are described in order to predict future movements (FISCHER, 1982).

Forecasting the generation of urban forestry waste is an important part of economic analysis and action planning, where there are five approaches to economic forecasting: exponential smoothing method; one-equation regression model; simultaneous equation regression model; autoregressive integrated moving average process (ARIMA) and autoregressive vector models (VAR) (GUJARATI; PORTER, 2008).

Initially, in order to obtain an economic cost of maintaining urban tree planting in a municipality such as Joao Pessoa, it is necessary to know its fraction in relation to total waste generation. In 2014, 5.98% of Joao Pessoa's MSW corresponded to biomass from urban tree maintenance. In Belo Horizonte (MG), tree planting waste accounted for 9.37% of the solid waste sent to processing plants. In Igaracy (PB), 7.10% of municipal solid waste is made up of biomass from urban tree planting. In the municipality of Goiana (PE), 1.41% of the MSW sent to processing plants was generated by urban tree maintenance services. In Jaboatao dos Guararapes (PE), 1.48 % of solid urban waste was generated by urban tree maintenance (SNSA, 2016).

With regard to the actual costs of disposing of urban tree waste, in the state of São Paulo, disposing of the material in a private landfill cost an average of R$ 68.00 per ton (MEIRA, 2010). In the city of Joao Pessoa/PB, costs are higher than the average for the state of Sao Paulo, corresponding to R$ 200.00 per ton, including transport, fuel, labor, landfill disposal and other operating costs (EMLUR, 2016).

A study carried out in 70 municipalities in the state of São Paulo found that there are no municipal regulations on the management of waste from the maintenance and removal of urban trees. It was clear that cities do not have established guidelines and responsibilities for tree-planting and pruning services (CAMILO et al., 2008).

With regard to the inadequate management of urban tree waste, poor disposal has an immediate impact on the environment and health, as well as contributing (through atmospheric emissions) to climate change. This contribution occurs during the decomposition stage. It forms toxic, asphyxiating and explosive gases and generates greenhouse gases (GHG), mainly methane (CH_4), as well as having the potential to contaminate surface and underground soil and water (GOLVEIA, 2012). It can be quantified by carrying out a Life Cycle Analysis (LCA).

Life Cycle Assessment (LCA) has become increasingly important in measuring the environmental impacts of activities and processes, as it stems from the awareness that improvements in a given process can induce secondary effects throughout the life cycle, which positively and/or negatively affect the environmental performance of goods and services (CARVALHO et al., 2015; FREIRE et al., 2015; PIRES et al., 2002).

The work was developed as follows: The first part consisted of a bibliographical review of urban tree planting (definition and benefits of urban tree planting; tree inventory; planning and tree planting, and; costs of maintaining tree planting), solid waste (definition of solid waste, and; National Solid Waste Plan (PNRS) and its applications in the municipality); biomass for energy purposes (definition of biomass and utilization technology for energy use), and Life Cycle Assessment (LCA) (application, importance and benefits).

The second part consisted of four articles. The first carried out a diagnosis and planning of urban tree maintenance in Joao Pessoa. A qualitative and quantitative assessment of the trees that make up the arborization.

The second article analyzed the carbon footprint associated with four scenarios (landfill, landfill with methane recovery, reuse and incineration) for urban tree waste in Joao Pessoa/PB. Evaluating, among the options analyzed, the most impactful and the most environmentally appropriate.

The third article evaluated the environmental impacts of three scenarios (electricity generation, heat generation and landfill disposal) for Joao Pessoa's urban forestry waste, as a strategy for inclusion in the Clean Development Mechanism (CDM).

The fourth and final article carried out a temporal analysis of the volume of wood from the urban afforestation of Joao Pessoa - PB, using a model from the ARIMA family.

1.1 OBJECTIVES

1.1.1 General objective

Carry out an environmental assessment and forecast of waste from Joao Pessoa's urban tree plantations as an energy alternative.

1.1.2 Specific objectives

a) Diagnosing and planning the maintenance of Joao Pessoa's urban tree planting;

b) Evaluate the environmental impacts and analyze four scenarios (landfill, landfill with methane recovery, reuse and incineration) for waste from the urban tree plantations of Joao Pessoa;

c) Evaluate the environmental impacts and analyze three scenarios (electricity generation, heat generation and landfill disposal) for Joao Pessoa's urban forestry waste, with a view to inclusion in the Clean Development Mechanism (CDM);

d) Analyze the predicted volume of waste from urban tree planting in Joao Pessoa.

1.2 THEORETICAL FRAMEWORK

In the theoretical framework, four themes were addressed for the development of the work: urban afforestation, solid waste, biomass for energy purposes and life cycle assessment, which are described below:

1.2.1Urban afforestation

The Egyptian tradition of creating green areas within human settlements was spread by ancient Asian and European civilizations, such as the Persians, Romans, Greeks and Eastern emperors, among others (LOBODA, 2005). This was one of the first actions that ancient civilizations took to keep in touch with nature, as villages became larger and more populous cities. At the same time, they began the first actions and activities that would result in the practice of urban afforestation.

In Brazil, the culture of afforesting cities is considered to be a recent activity, compared to cities in European countries, starting around 120 years ago (DANTAS; SOUZA, 2004). It is understandable that the culture of afforesting cities in Brazil is considered recent, as we inherited this habit and tradition from Europeans during the Portuguese and Spanish colonization of Latin America.

Urban afforestation refers to a pattern of tree distribution in an urban area, constituting a forest (ROSSETTI et al., 2010). It is defined as the arboreal vegetation elements that a city has, or even a set of arboreal or cultivated vegetation that a city has, in private and public areas (SANTOS et al., 2008; SANCHOTENE, 1994).

The implementation of urban forests must be carried out in a technical manner, being idealized, planned and/or assisted by the public authorities, providing better conditions for the development and conservation of trees (RECIFE, 2013).

Public authorities have the technical and financial resources to plan and implement tree planting in a technical and safe manner, avoiding future damage to the population and the city's structure. When implemented in accordance with technical standards, the chances of vegetation making a positive contribution to the city's dynamics increase.

Urban vegetation contributes to the formation of a healthy and comfortable physical environment through environmental services, such as: regulating and purifying air, soil and water; mitigating heat and stabilizing the climate; providing shelter and food for fauna; protecting the soil from erosion, and; reducing levels of human stress; among others (VALE et al., 2011; MELO; SEVERO, 2010).

In order to optimize the benefits that urban afforestation provides to society, it is necessary to know the characteristics and plant heritage. This will provide public authorities with information to carry out corrective actions and plan future activities.

Knowledge of plant heritage in urban areas is achieved through an arboricultural inventory. It consists of collecting information about existing specimens and the places where they are located, in order to assess their condition and structural characteristics. The inventory also identifies trees in need of intervention and helps assess the cost of afforestation (CEMIG, 2011).

The pioneers in urban tree inventories carried out in Brazil were Biondi (1985) and Milano (1987), both with the aim of characterizing the tree composition of the municipalities of Recife/PE and Curitiba/PR, respectively.

The study by Biondi (1985) analyzed the techniques used in tree planting and determined which species were the most suitable. The author pointed out that urban tree planting needed to be improved, starting with better planning, through an evaluation and analysis of urban tree planting, in order to obtain an adequate basis for planning.

Milano (1987) evaluated the treatment and management needs of street trees, as a space dependent on urban planning. He concluded that it was possible to minimize maintenance costs and improve the quality of tree planting by selecting species in terms of shape, size and available space, improving the quality of seedlings and maintenance techniques.

The arboricultural inventory is an important tool for the management of urban trees in any city, as it provides guiding information for management actions. It identifies sites with the potential to receive trees, as well as their situation, which influences and corroborates actions aimed at their maintenance and better management of public money.

Lack of planning during the implementation phase of urban tree planting can result in significant impacts on the city's structure, such as: disruption of high-voltage wires; clogging of sewage systems and gutters; and cracks in sidewalks, among others (NUNES et al., 2013). This results in the need for tree maintenance or the removal (or replacement) of trees, generating a significant amount of waste (BARROS, 2013).

For comparative purposes, in the municipality of Irati (PR), 32.36% of the trees needed cultural treatment, due to conflicts with urban equipment (SCHALLENBERGER et al., 2010). In the city of Quirinópolis (GO), 56.60% of the trees in the Promissao neighborhood and 50% in the Pedro Cardoso neighborhood needed light or heavy pruning (BATISTEL et al., 2009).

The maintenance and removal of urban trees has costs, which are borne by the municipality or the energy concessionaire. In the city of Curitiba (PR), the total cost of tree maintenance carried out by the municipality was R$34.31/small tree and R$183.50/large tree, with cycles every 12 years. For the energy concessionaire, the costs vary between R$ 29.73 and R$ 108.32/tree, depending on the size, with a five-year cycle (LEAL et al., 2008).

In the cities of Santa Monica and Modesto, California (USA), the cost of tree pruning varies according to the periodicity (pruning cycle), US$ 90.00 from one to three years, and US$ 69.00 with cycles of three to four years (McPHERSON; SIMPSON, 2002).

The total cost of removing trees (aerial part and stump) in the city of Curitiba varied between R$ 94.31 and R$ 243.50, depending on the size of the tree, and the removal of the stump was considered to cost an average of R$ 60.00 (LEAL et al., 2008). In the cities of Modesto and Santa Monica, California (USA), the average costs for total tree removal ranged from US$ 264.00 and US$ 396.00, respectively (McPHERSON; SIMPSN, 2002).

Much of the maintenance of urban tree planting is carried out by municipalities, and even though it is necessary to carry it out to avoid problems for third parties, it can consume significant financial resources for small municipalities. These costs could be reduced, and in some cases eliminated, when planted in suitable locations, according to the phenotypical characteristics of the plant and the structure surrounding the site where the tree will be planted.

In urban afforestation, the planting of trees on sidewalks must reconcile two objectives: diversity of species and uniformity of composition, with a view to the environmental quality of afforestation plans, the practical quality of management and hostile environments (BOBROWSKI; BIONDI, 2012).

Hostile environments represent a major stumbling block for plant development in urban areas, including: atmospheric pollution; soil contamination; lack of sufficient space; and man-made mechanical injuries. In addition to interference from inadequate paving, bottlenecks in flowerbeds and conflicts between roots and gas, water and other utility pipes (ROSSETTI; PELLEGRINO; TAVARES, 2010).

These diversities in urban areas indicate the need for public authorities to act in the implementation and knowledge of tree heritage. Possible actions to encourage the development and implementation of urban tree planting should be devised as public policy, with the aim of providing better conditions for tree growth.

1.2.2Solid waste

Solid waste (SW) is a delicate issue for most municipalities, as it is generated daily and in large volumes. Many Brazilian municipalities do not have an adequate system for managing their waste, disposing of it in an irregular manner that has an impact on society (FIGUEIREDO, 2012). There is a need to create legal instruments and technical feasibility for their proper disposal without causing environmental impacts.

In 2010, Brazil created the National Solid Waste Policy (PNRS), instituted by Law No. 12.305/2012 (BRASIL, 2010a), regulated by Decree No. 7.404/2010 (BRASIL, 2010b), which sets out the principles, objectives and guidelines related to integrated waste management.

Decree No. 7.404/2010 (BRASIL, 2010b), in addition to regulating the PNRS, also created the Interministerial Committee for the National Solid Waste Policy

(CIPNRS). The purpose of this Committee is to support the structuring of the National Solid Waste Policy by coordinating government bodies and entities (BRASIL, 2010b).

The CIPNRS is made up of representatives from the Ministry of the Environment (which coordinates it); the Civil House of the Presidency of the Republic; the Ministry of Cities; the Ministry of Social Development and Fight against Hunger; the Ministry of Health; the Ministry of Mines and Energy; the Ministry of Finance; Ministry of Planning, Budget and Management; Ministry of Development, Industry and Foreign Trade; Ministry of Agriculture, Livestock and Supply; Ministry of Science and Technology; and the Secretariat for Institutional Relations of the Presidency of the Republic (BRASIL, 2010b).

The legal creation of these regulations was an important step in encouraging the development of actions aimed at resolving conflicts between municipalities and solid waste. With regard to the creation of the inter-ministerial committee by the PNRS, this has broadened discussions on solid waste, as it encompasses various sectors of government activity, but has left out representatives of civil society and scholars in the field.

There are several definitions of solid waste in the literature, two of which stand out for their detail, legal and technical application. The one described by the PNRS (BRASIL, 2010a), which defines solid waste as:

"Art. 3 ...:

(...)

XVI - solid waste: discarded material, substance, object or good resulting from human activities in society, the final destination of which is solid or semi-solid, as well as gases contained in containers and liquids whose particularities make their discharge into the public sewage system or bodies of water unfeasible, or require solutions that are technically or economically unfeasible in view of the best available technology;

(...)" (BRASIL, 2010a).

And according to the Brazilian Association of Technical Standards, waste

solids are (ABNT, 2004):

"Solid waste; solid and semi-solid waste resulting from industrial, domestic, hospital, commercial, agricultural, service and sweeping activities. This definition includes sludge from water treatment systems, those generated in pollution control equipment and installations, as well as certain liquids whose particularities make it impossible to discharge them into the public sewage system or bodies of water, or require technical and economically unfeasible solutions in view of the best available technology." (ABNT, 2004)

Both definitions converge on the fact that solid waste is material resulting from human activities, including actions in the production process of products and services, whether durable or not. By definition of the PNRS, gases contained in containers are also considered solid waste.

Another action highlighted in the PNRS was the preparation of state and municipal solid waste plans. Generally speaking, the State Solid Waste Plans are valid for an indefinite period, covering the entire territory and with an action horizon of 20 years, with four-yearly updates, covering at least:

"Art. 17 ....

I - a diagnosis, including the identification of the main waste flows in the state and their socio-economic and environmental impacts;

II - proposing scenarios;

III - targets for reduction, reuse, recycling, among others, with a view to reducing the amount of waste and refuse sent for environmentally appropriate final disposal;

IV - targets for the energetic use of gases generated in solid waste final disposal units;

V - targets for the elimination and recovery of waste, associated with the social inclusion and economic emancipation of collectors of reusable and recyclable materials;

VI - programs, projects and actions to meet the targets set;

VII - technical standards and conditions for access to state resources, for obtaining its endorsement or for access to resources administered directly or indirectly by a state entity, when intended for actions and programs of interest to solid waste;

VIII - measures to encourage and make feasible the consortium or shared management of solid waste;

IX - guidelines for planning and other solid waste management activities in metropolitan regions, urban agglomerations and micro-regions;

X - norms and guidelines for the final disposal of tailings and, where applicable, waste, respecting the provisions established at national level;

XI - provision, in accordance with other territorial planning instruments, especially ecological-economic zoning and coastal zoning, for:

a) favorable areas for the location of solid waste treatment plants or final disposal of tailings;

b) Degraded areas due to inadequate disposal of solid waste or tailings to be subject to environmental recovery;

XII - means to be used for control and inspection, at state level, of its implementation and operation, ensuring social control." (BRASIL, 2010a).

In Paraiba, the State Solid Waste Plan (PERS) addresses the issue of eradicating garbage and disposing of waste in sanitary landfills; recovering areas degraded by solid waste; using the gases generated in sanitary landfills for energy and recovering waste (PARAIBA, 2014).

For Municipal Solid Waste (MSW), the PERS provides for the implementation and coverage of selective collection in the municipalities, waste pickers organized in associations and cooperatives, treatment of the organic fractions of MSW implemented and in operation, and the sorting and processing of recyclable waste. The PERS indicates guidelines for reducing solid waste disposed of in landfills, integrating cooperatives and associations of recyclable material collectors and encouraging the practice of reusing and recycling solid waste (PARAIBA, 2014).

Among the public service management strategies for urban cleaning and solid waste management is support for the preparation, implementation and review of Integrated Municipal and Intermunicipal Solid Waste Management Plans (PARAIBA, 2014).

The disciplining of solid waste management at state level has provided municipalities with guidance on possible management actions and proper disposal, especially small ones, as required by law and technically acceptable.

The PNRS requires municipalities to develop a Municipal Plan for the Integrated Management of Solid Waste (PMGIRS), and it is the municipality's responsibility to prepare and implement it. The PMGIRS for the municipality of Joao Pessoa was established in 2014 and consists of two volumes, one presenting the diagnosis and the other the prognosis (EMLUR, 2014a).

The prognosis of the PMGIRS corresponds to the guidelines, strategies, goals, programs and projects. Among the general guidelines are the non-generation and reduction of Municipal Solid Waste (MSW), reuse and recycling, and treatment of MSW with appropriate and economically viable environmental technology, among others (EMLUR, 2014b).

These guidelines follow those of the national and state plans, but are adapted to local specificities. Adapting national and state plans to local characteristics avoids errors in the implementation of waste management plans and programs, as well as the execution of unnecessary or inappropriate services.

The objectives of the PMGIRS include: encouraging the development of actions that rethink consumption and the generation of MSW; practicing reverse logistics in commercial establishments and empty pesticide packaging; encouraging the management of construction waste (RCC); and requiring the preparation and implementation of a health waste management plan (EMLUR, 2014b).

In order to achieve the objectives of the PMGIRS, actions are planned, such as promoting studies and research into technological alternatives for treating MSW, processing urban forestry waste, and setting up new sorting units (EMLUR, 2014b).

Once the plan has been implemented and the objectives achieved, it is hoped to improve the cleanliness of the municipality, eliminate dumping grounds, provide proper disposal of construction waste, contribute to the conservation of water resources in the municipality, among other things (EMLUR, 2014b).

The management of solid waste at municipal level is of fundamental importance in order to ensure that it is disposed of properly, directly at the source of generation. Raising awareness and sensitizing the population is a tool envisaged in all governmental spheres, and this is the beginning of a change in behavior and attitude in the largest generating source of solid waste.

1.2.3Biomass for energy purposes

Biomass can be defined as the organic mass produced per unit area, expressed in dry matter weight, wet matter weight or carbon weight. Biomass can be made up of a multitude of materials and can be used for a variety of purposes (SOARES et al., 2011).

Biomass energy can be defined as a renewable resource derived from organic material, of animal or plant origin, capable of producing energy (BRACMORT, 2015; SANTOS, 2012). It has been used since the beginning of civilization, allowing primitive man to evolve his living habits by producing light and heat (PRETO; MORTOZA, 2010).

Biomass energy is one of the most widely used renewable energy sources in the world, due to its accessibility and availability (BRAND et al., 2014; MACEDO, et al., 2014). It is used *in natura* (firewood), charcoal or in the use of forestry waste (tree maintenance waste, branches and leaves) or industrial waste (wood chips, sawmill and laminate waste, briquettes) (VIEIRA et al., 2014; MOREIRA, 2011; COUTO et al., 2004).

Forest biomass for energy purposes has the greatest potential for use in terms of nature, origin, conversion technology and energy products, with great technology for conversion processes, including direct combustion, briquettes and pellets (SAWIN, et al., 2012; BRASIL, 2007), making it an alternative for diversifying the energy matrix.

Due to its potential as an energy and conversion source, biomass plays a significant role in the global and national energy matrix. This has been the result of new research, investment in technological innovation to convert biomass into energy and the optimization of existing energy systems.

In 2014, biomass accounted for 10% of the world's primary energy supply, most of which was consumed for cooking and heating. In 2012, 1.5% (corresponding to 370 TWh) of the world's electricity came from this energy source (IEA, 2014). In Brazil, 23.8% of the domestic energy supply was generated from biomass, with 15.7% coming from sugar cane bagasse and 8.1% from firewood and charcoal (EPE, 2015).

In some regions of the planet, the use of biomass in the form of heat is greater than electricity, due to the technology, equipment and initial financial resources required for such conversion. The transformation of wood into other products is a viable alternative when there is a need to obtain a product capable of releasing a large amount of energy.

Briquettes are one of the technologies used to increase energy output by transforming biomass. It is the result of compacting lignocellulosic waste using a mechanical press with pressure, with or without the use of temperature, with or without the aid of binders (DIAS, 2012; ROSARIO, 2011; SANTOS et al., 2011). Its main characteristic is the high amount of energy released during burning, compared to other *in natura* products.

In order to make better use of both briquettes and *raw* firewood, it is necessary to study their energy properties, the most important of which is calorific value (ELOY, 2015; BRAND, 2010).

The calorific value can be defined as the amount of heat or thermal energy released during the total burning of a given amount of fuel, usually expressed in kcal per kg of fuel (SILVA, 2012a; CINTRA, 2009; JARA, 1989).

Calorific value is an important parameter for analyzing feasibility and helps make decisions about the use of biomass as energy, and which conversion process to use. Among the conversion processes, gasification has attracted the attention of researchers, as it is a technique that requires more specific equipment and generates a gas as its final product.

The process of gasifying biomass waste involves 4 stages: Drying, pyrolysis (devolatilization or thermal decomposition), combustion or reaction of the oxygen material, and finally, gasification (RUMÂO, 2013).

This process of using biomass for energy occurs through the use of "chemically stored" energy converted into heat through combustion, which is a thermochemical conversion. This thermochemical process can also include direct burning and pyrolysis (BRASIL, 2007). Other ways of using energy are biochemical conversion, which includes anaerobic digestion and fermentation, distillation and hydrolysis, and physical-chemical conversion, which involves the compression and extraction of oils (BRASIL, 2007).

The choice of technology for converting biomass into energy must take into account the purpose, analyzing what end product you want to obtain. Due to its versatility in terms of the end product, biomass has its place in the energy matrix, even with the increase in other fuel sources. Continued research into optimizing the energy generated by biomass could be a way of increasing its use and diversifying the national energy matrix.

1.2.4Life Cycle Assessment

Life cycle assessment (LCA) emerged with the increase in public awareness of the need for environmental protection. It is a widely applied methodology for carrying out an environmental assessment in the industrial production chain (goods) and service provision, not only in manufacturing, but also with a focus on where the impacts are really significant. It is widely used by various specialists, mainly in the scientific and engineering fields around the world (ABNT, 2014; CAVALETT, 2008).

The development of this tool has brought about a change in the relationship between consumers and producers (services or industries). It has enabled the population to have another tool, in addition to the amount to be paid for the product or service, to help them select the product to be purchased and rethink what its final destination will be.

LCA has become increasingly important in measuring the environmental impacts of activities and processes, as it stems from the awareness that improvements in a given process can induce secondary effects throughout the life cycle. This positively and/or negatively affects the environmental performance of goods and services (CARVALHO et al., 2015; FREIRE et al., 2015; PIRES et al., 2002). A life cycle is understood as the necessary stages for a product or service to be developed or designed, fulfill its respective function and reach the stage of disposal, recycling or reuse.

LCA is carried out using software such as Simapro® and Gabi®. These have a number of advantages, including: ease and practicality of use; speed in obtaining results due to the reduction in evaluation time; standardization of the databases used; standardization of the way results are presented and the simplifications assumed, as well as ease in comparing the results obtained with the literature (CAVALETT, 2008).

However, this *software* has some limitations, one of which is the database in the American and European systems, which in some situations are quite different from the reality in Brazil. In addition, due to the need to standardize the use of resources, in some cases the *software* makes excessive simplifications, making it impossible to describe the appropriate systems and some interesting characteristics are lost (CAVALETT, 2008).

The use of this *software* makes it easier and possible to obtain more accurate results of emissions at each stage of production and service. This contributes directly to the development of actions and plans aimed at optimizing production and reducing GHG emissions.

Even with some limitations in the use of software, LCA is an important tool because it makes it possible to clearly and objectively address and find solutions for managing natural resources, optimizing production systems, defining parameters for environmental labelling, among others (GATTI, 2002).

LCA has a wide range of applications in the private sector, NGOs and the public sector. In the public sphere, its use has the potential to generate subsidies for environmental public policies (SILVA, 2012 b). In some European countries, LCA has been used to develop Integrated Product Policies (IPP), aimed at encouraging greener consumption, taxing more polluting products and creating environmental labeling in companies (HAUSCHILD; POTTING, 2005).

This application in European countries proves its effectiveness and viability for public authorities, helping to create environmental public policies that are more appropriate to the reality of each country. Even though it is sometimes a complex and expensive system, it is possible to adapt, apply and use it in Brazilian municipalities, adjusting it to local peculiarities.

With regard to the issues of costs and complexity in LCA, these can be dealt with by developing a database on the production chain of a wide variety of consumer goods. These include databases on various materials (ceramics, metals, composites, etc.), energy (electricity, heat, etc.), transportation (road, air, sea, etc.) and waste (gaseous emissions, liquid effluents and solid waste), helping to reduce costs and shorten the LCA process (SILVA, 2012 b).

The treatments available in LCA today are so wide-ranging that they make it possible to carry out actions in various productive and service sectors, from the manufacturing industry to mining. The forestry sector is an industrial segment that uses natural resources directly, and it is extremely important to maintain a balance between product generation and the final waste generated.

According to the LCA literature in the forestry area, there is a demonstration of applicability in breadth and diversity, with the works of Brugnara (2001), Mastella (2002); Silva (2012 B); Haaren et al. (2010); Morris et al. (2011); Zhang (2012); CANADA (2014); Reichert and Mendes (2014);

Brugnara (2001) analyzed the energy consumption and CO_2 emissions in the production chain of three wood supply systems for use in residential construction: (1) wood from managed native forests; (2) wood from unmanaged native forests and (3) wood from eucalyptus plantations. The result has the potential to serve as a tool for consumers to make decisions about the product. They have the option of choosing between the possible origins of the wood, which product to choose, taking into account the lower or higher carbon emissions.

Mastella (2002) compared the production processes of ceramic and concrete blocks for structural masonry. Silva (2012) carried out an LCA of the production of a *medium density particle board* (MDP) made from wood particles and synthetic adhesive. Haaren et al. (2010) compared the environmental impacts of composting yard waste with landfill disposal.

The work of Morris et al. (2011) reviewed various final disposal scenarios for *leaf and tree waste* from the municipality of Red Deer (Canada), concluding that landfill with gas recovery for energy was preferable to incineration with electricity generation (considering climate change, human carcinogens and toxicity to the ecosystem).

When comparing disposal for composting and biogas production in landfills, Zhang (2012) found that biogas production can be competitive in terms of CO_2 -eq emissions compared to composting if the purification of the biogas can be improved by reducing the electricity consumption required.

The study by *The Leaf and Yard Waste Diversion Technical Committee* (CANADA, 2014) indicated that there are better options for urban tree waste than landfill (composting, for example), mainly for economic reasons. Including not using space (land use), and for environmental reasons (reduction of GHG emissions).

Following the example of the studies presented, there may be better options for disposing of biomass waste from the maintenance of urban trees in Joao Pessoa. Bearing in mind that they reported other examples of more environmentally beneficial disposal, apart from landfill. It is worth noting that the other disposal options may be even more economically viable, as well as environmentally, and have the potential to be applied as a public policy in the city.

Reichert and Mendes (2014) studied LCA and supported its use in decision-making in integrated and sustainable management of solid urban waste, considering the targets for reducing the disposal of organic waste in landfills set by the PNRS: 60% for the southern region by 2031.

Therefore, the application of LCA to support municipal decision-making and the management of GHG emissions from products and services has proved to be technically feasible in studies similar to the one carried out in this research, making it an indispensable tool to be taken into account before decisions are made.

1.2.5Time Series

A time series, or historical series, is defined as a sequence of data obtained at a certain regular interval of time. With a time series, you want to model the phenomenon being studied, describe the behavior of the series, make estimates and evaluate which factors influence its behavior. It has applications in various areas: insurance systems, industries, economics, medicine, epidemiology, communications, reliability, methodological phenomena, forestry, among others (ANDRADE, 2013; LATORRE; CARDOSO, 2001).

The analysis of a time series is concerned with deriving knowledge about recent and past movements, in order to predict results and recognize factors that are relevant to the object of study. Forecasting is a useful tool that enables more efficient actions and better planning, supporting decision-making (ANTUNES; CARDOSO, 2015).

Time series are widely applicable, covering various segments of the production chain. However, there are several models for analyzing time series, and it is necessary to select the most suitable model, depending on the object of study and objective. When analyzing a time series where there is no trend or seasonality, autoregressive models (AR) or models incorporating moving averages (ARIMA) can be used. When there is a trend, autoregressive integrated moving average models (ARIMA) and SARIMA models can be used to incorporate seasonality (LATORRE, CARDOSO, 2001).

The modeling of time series using the ARIMA method is well known and used in the analysis of non-stationary series, and was proposed by Box and Jenkins (1970). It is an extension of the ARMA parametric linear models (AGUIRRE, 2007; CAMPOS, 2008).

Some areas use time series to carry out specific studies and analyze performance, understanding future behaviors or trends. In the health area, Amâncio and

Nascimento (2012) used time series to estimate the risk of children being hospitalized for asthma after exposure to pollutants in Sao José dos Campos (SP).

In the agricultural sector, Bressan (2004) analyzed the applicability of time series models as a decision-making tool for buying and selling futures contracts on the BM&F. The results indicated the potential of using the MDL and ARIMA forecasting models as a decision-making tool for trading in contracts for dates close to maturity.

In the forestry sector, Floriano *et al.* (2006) used time series to develop height growth equations that better adapted to population data and individual selection criteria in a *Pinus elliottii population*.

The time series of the price of natural rubber was studied by Coelho Junior et al. (2009), who analyzed the behavior of prices on the international market from January 1982 to December 2006. Among the conclusions, the model that best fitted the series studied was AR (1) for a GARCH (1,1). And Soares *et al.* (2008) developed a model to estimate the domestic market price between January 1999 and September 2008. The ARIMA (2,1,1) model was the most suitable for forecasting the price of rubber in Brazil.

Coelho Junior et al (2006) analyzed the longitudinal series of charcoal prices in the state of Minas Gerais, concluding that the SARIMA model (2,0,1)(0,1,1) provided the best fit and was parsimonious. Soares et al (2010) developed a model to predict the price of *Eucalyptus spp.* standing timber in Itapeva (SP) and Bauru (SP). The ARIMA (0,1,4) and ARIMA (2,1,2,) models were the most suitable for forecasting in the two cities, respectively.

Almeida et al (2009) studied a model for predicting the prices paid for plywood exports from Paraná, where graphical and statistical analysis indicated that the ARIMA (1,1,3) model was the best fit for the plywood price series.

2. ARTICLE 1 - DIAGNOSIS AND PLANNING OF URBAN TREE PLANTING IN JOAO PESSOA

YURI ROMMEL VIEIRA ARAÙJO; ZAYNE CHRISTINA GONÇALVES MOREIRA; LUIZ MOREIRA COELHO JUNIOR.

SUMMARY

This study diagnosed and planned a strategy for urban tree planting in Joao Pessoa. For the diagnosis, an inventory of urban trees was carried out by means of simple random sampling, using streets and avenues as sampling units. From this sampling, the linear meter density (LMD), the Shannon-Weaver diversity index (H') and equability, using the Pielou equation, were analyzed. The main results were: 101 species distributed in 32 families were identified; Fabaceae was the most representative and Moraceae had the largest number of individuals; of the species identified, 57.43% were exotic and 42.57% native, with 78.72% of the individuals being exotic (2,420 specimens) and 21.28% native (654 specimens); *Ficus benjamina* L., with 34.29%, was the most abundant species of the total sampled; of the individuals that presented problems, 76.57% had conflicts with public equipment, 23.25% with insects, 7.23% with pathogens, and 27.94% with physical damage; the average DML was 37.66 individuals/km of sidewalk; H' was 2.96 nats/ind and equability was 0.37. The urban forest is unevenly distributed throughout the municipality, so an urban tree management plan needs to be formulated with a tree monitoring program, focusing on individuals with phytopathological problems, physical damage and conflicts with urban equipment, by registering the trees in the different sectors/neighbourhoods of the city.

Keywords: Tree inventory, management plan, urban afforestation.

2.1 INTRODUCTION

Urban centers are places where a large part of the population lives together. They are constantly being modernized and/or altered, resulting in an imbalance between man and vegetation (ANGELIS et al., 2007). One of the tools for maintaining contact with nature is the afforestation of streets and avenues, which in Brazil can be considered a recent activity when compared to European countries (DANTAS; SOUZA, 2004).

Urban afforestation refers to a pattern of tree distribution in an urban territory, on public roads and other areas free of buildings. The success of afforestation depends on correct and careful planning, taking into account the factors that influence species selection, seedling production and implementation (BOBROWSKI; BIONDI, 2012).

Poor planning of tree planting during the implementation phase of a project can result in significant impacts on the urban structure. For example, the breakage of power lines resulting in interruptions to the electricity supply, clogging of sewage systems and gutters, cracks in sidewalks, among others (NUNES et al., 2013).

To correct an unstructured and disorganized arborization, it is necessary to prune or remove (or replace) branches and trees (BARROS, 2013). Pruning is the elimination of branches, twigs, inflorescences or foliage in order to make it compatible with the existing physical space or to promote the proper development of the plant, and can be formative, conductive, cleaning, corrective, adequate, lifting and emergency (RECIFE, 2013; SÂO PAULO, 2015).

In the municipality of Irati (PR), 32.36% of the trees needed cultural treatment due to urban conflicts (SCHALLENBERGER et al., 2010). In the city of Quirinópolis (GO), 56.60% of the trees in the Promissao neighborhood and 50% in the Pedro Cardoso neighborhood needed light or heavy pruning (BATISTEL, 2009). In the municipality of Joao Pessoa/PB, in 2012, the city council pruned 9,200 trees and replaced 400, through the implementation of the Programmed Pruning program (JOÂO PESSOA, 2012).

The development of monitoring and maintenance programs is important for urban tree planning. Silva Filho et al. (2002) showed that in the city of Jaboticabal (SP) it is feasible to create a database for registering, evaluating and managing trees to help with corrective actions and tree management. Another way is through the geographic information system, indicating the location, maintenance history, conflicts with public equipment, electricity and water networks, sewage, among others (OLIVEIRA FILHO; SILVA, 2010).

A forest management plan begins with an arboricultural inventory. This corresponds to a diagnosis of the tree situation by collecting information on planting sites, identifying maintenance, repairs or removal, among others (BENATTI et al., 2012; CEMIG, 2011). The inventory can be carried out using three methodologies: partial inventory (square, neighborhood); complete inventory (tree census of parks, streets or green areas); and sample inventory (BOBROWSKI, 2010).

Inventory sampling can be done using simple or casual random sampling, systematic sampling or restricted random sampling. Among the sampling methods, random sampling is carried out by drawing sample units (rectangular or quadrangular plots, blocks or streets) at random or at defined intervals. This model makes it possible to obtain the average value of the variables analyzed and to estimate the accuracy of these averages (ROSSETTI, et al., 2010). This sampling method is a relatively easy way of getting to know the structure of urban forests and estimating the values of environmental services, where the accuracy and cost of the estimate depend on the size of the population and the sampling units (NOWAK et al., 2008).

Periotto et al. (2016), inventoried the urban forestry of the city of Medianeira/PR by means of random sampling in order to select the sampling units of the object of study. In the city of Huambo/Angola, sampling took place in the city's main streets and determined sampling sufficiency using the species richness curve (QUISSINDO et al., 2016). In Boa Vista - RR, urban tree planting was analyzed through a census of the municipality's streets (LIMA NETO et al. 2016). For the inventory in the Vila Yolanda neighborhood, in the municipality of Foz de Iguaçu/PR, eight streets in the neighborhood were randomly selected to collect data on all the trees on the streets (TOSCAN et al., 2010).

The municipality of Joao Pessoa does not yet have an inventory of urban trees and its systematic tree conservation planning. Bearing in mind the importance of information on urban vegetation, this study diagnosed and planned the maintenance of Joao Pessoa's urban trees.

2.2 MATERIALS AND METHODS

2.2.1 Object of study

The study was carried out in the municipality of Joao Pessoa, capital of the state of Paraiba, located in the Zona da Mata Paraibana. The municipality has a population of 801,718 inhabitants in 2016, and an area of 211.475 km^2 of territory, resulting in a demographic density of 3,421.28 hab/km2 (IBGE, 2016).

According to the Koppen-Geiger climate classification, the municipality is located in the 3dth climate, corresponding to the Mediterranean or Northeastern sub-dry climate, in the AS' type climate band, a hot and humid climate (Rainy Tropical - class A). The months of May and July see the highest rainfall, with an average of 1,896 mm/year. Daily maximum and minimum temperatures range from 30 to 21 °C, and relative humidity varies between 73% and 82% (SOUZA et al., 2016).

The city has low altitudes in relation to sea level and remnants of Atlantic Forest, cerrado and tabuleiro, forming fragments throughout the city, mainly in the river valleys (SANTOS; SANTOS, 2013).

2.2.2 Tree inventory and data analysis

2.2.2.1 Sectorization of the study area

For the tree inventory of the city of Joao Pessoa/PB, the simple random sampling method was used, using streets and avenues as sampling units. To carry out the fieldwork and data collection, the urban perimeter was divided into 4 sectors, as shown in Figure 2.1. The data collection period was from July to November 2015.

For each sector, the neighborhoods, number of streets, nomenclature and linear extension of each street were identified with the help of the cadastral form of the Municipal Planning Department (JOÂO PESSOA, 2015). Green areas, squares, parks and private gardens were not considered.

Sector A: Aeroclube, Altiplano Cabo Branco, Bairro dos Estados, Bairros dos Ipês, Bancàrios, Bessa, Brisamar, Cabo Branco, Castelo Branco, Jardim Oceania, Joao Agripino, Manaira, Miramar, Pedro Gondim, Ponta do Sol, Sao José, Tambaù, Tambauzinho and Torre.

Sector B: Agua Fria, Anatólia, Barra de Gramame, Cidade dos Colibris, Costa do Sol, Cuià, Jardim Cidade Universitària, Jardim Sao Paulo, José Américo, Mangabeira, Muçumagro, Paratibe, Penha, Planalto da Boa

Hope and Valour.

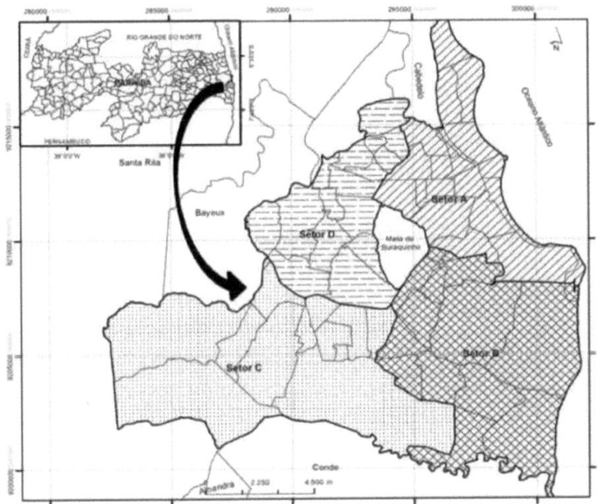

Figure 2.1. Delimitation of the sectors for the Tree Inventory of the City of Joao Pessoa/PB.

Map: Arinaldo Inacio.

Sector C: the neighborhoods of Industrias, Costa e Silva, Distrito Industrial, Ernani Sàtiro, Ernesto Geisel, Funcionàrios, Gramame, Grotao, Jardim Veneza, Mumbaba and Mussuré.

Sector D: Alto do Céu, Alto do Mateus, Centro, Cristo Redentor, Cruz das Armas, Ilha do Bispo, Jaguaribe, Mandacarù, Oitizeiro, Padre Zé, Rogér, Tambià, Treze de Maio, Trincheiras, Varadouro and Varjao.

2.2.2.2 Selection of sampling units, Estimates of the optimum number of sampling units and Linear Meter Density (LMD)

The streets were selected using systematic sampling. For each sector, the streets were sorted alphabetically. The first 10 streets were randomly drawn and every 10th street was systematically selected.

A pilot tree inventory was carried out in 20 streets in order to obtain a sufficient sample. Biondi (1985) was used to obtain the number of streets. Based on the pilot inventory, 241 streets were inventoried (3.86% of the total).

The linear meter density (LMD) of urban trees was calculated by dividing the total number of individuals recorded in the sampling interval by the total length of the public sidewalk sampled.

The Shannon-Weaver index (H') verified population diversity, obtained from the expression:

$$H' = \frac{\left[N \ln(N) - \sum_{i=n}^{S} n_i \ln(n_i)\right]}{N}$$

Where:

H' = Shannon-Weaver index;

n_i = Number of individuals sampled of the i-th species;

N = total number of individuals sampled;

S = total number of species sampled;

ln= Neperian base logarithm.

The Pielou equation was obtained from the expression:

$$J = \frac{H'}{H_{max}}$$

In which,

H_{max}= ln (S);

S = total number of species sampled;

J = Pielou's Equability;

H' = Shannon-Weaver diversity index.

2.2.2.3 Field data collection and analysis

In the selected streets, a tree census was carried out of individuals with a chest height circumference (HC) ≥ 15 cm, on the sidewalks and central flowerbeds, excluding squares, parks and green areas, HC at 1.30 m from the ground, and total height of the individual (HT). The common name of the plants was collected in the field for later identification of the scientific name,

The botanical identification of the species was carried out using vegetative and reproductive characteristics (rhytidome, leaves, flowers and/or fruit). For trees that were not identified *on site*, samples of fertile structures were collected, when possible, for identification by a specialist or for identification by comparison with already identified samples deposited in the Lauro Pires Xavier Herbarium at the Federal University of Paraiba.

The classification of botanical families was based on the *Angiosperm Phylogeny Group* IV (APG IV) (APG, 2016). The scientific names of the species and families were checked against the Brazilian Flora Species List (REFLORA, 2017) and the *Missouri Botanical Garden*'s tropical plant database (TROPICOS, 2016).

The species were also classified according to the phytogeography of Brazil (native and exotic species). References were Lorenzi et al. (2003), Lorenzi et al. (2010) and the List of Brazilian Flora Species (REFLORA, 2017).

Phytosanitary aspects were also observed, in terms of the presence or absence of insect pests (termites, leaf-cutting ants, caterpillars, beetles, bedbugs, fleas, mites, etc.) and the presence or absence of pathogenic diseases (microorganisms). Physical damage to the trunk (defects and physical damage, peeling and ringing, trunk object and cuts in the trunk other than maintenance), and conflicts with public equipment (electricity and telephone network, sewage pipes, *water* supply, *and* lifting or breaking of sidewalks and water lines).

2.2.3João Pessoa's urban forestry improvement plan

A program of corrective and preventive actions for urban tree planting and the management of urban tree maintenance waste was drawn up. The results of the tree inventory showed the need for action and potential risks in the urban environment. The percentage of individuals of each of the 30 most representative species was taken into account in order to identify those that are most adaptable and resistant to pathogens in the city.

2.3 RESULTS AND DISCUSSION

2.3.1 Floristic survey

The tree inventory in the 241 streets sampled counted 3,074 individuals belonging to 32 families, distributed in 101 species, of which 8 species were identified to genus level. Of the total found, 195 were palm trees (6.34% of the population sampled) and 2,879 were trees (93.66% of the population sampled), as shown in Table 2.1.

The number of species found in a city varies according to anthropogenic, cultural and geographical factors. Climatic factors directly influence the choice of species used on access roads. It is not always the case that the species chosen to be planted voluntarily by the population on the sidewalks of their homes in order to protect them from the sun is a species suitable for this purpose. The urban floristic richness found in Joao Pessoa is average when compared to other cities studied in Brazil. In the south of the country, Lindenmaier; Souza (2014) found 101 species in Cachoeira do Sul/RS and Salvi et al. (2011) cataloged 61 species in Porto Alegre/RS. In the southeast, the roadside arborization of Sorocaba/SP was composed of 203 species (CARDOSO-LEITE et al., 2014), while in Jaù/SP 75 species were found in the urban arborization (SOUZA et al., 2004).

In Joao Pessoa, according to the tree inventory, the Fabaceae family was the most representative, with 25 species (24.75 %). Followed by Bignoniaceae with 11 species (10.89 %) and Arecaceae with 10 (9.90 %). Together, these families account for around 50% of the species identified in the study. However, the predominance of the Fabaceae in urban tree planting is not exclusive to Joao Pessoa because according to Hueck (1972), the Fabaceae is the richest and most abundant family in South America. Other Brazilian cities studied show this to be true: In Timon/MA, Morais and Machado (2013) found 23 species (26.74%) in the trees. The same was found in Itapuranga/GO, with 11 existing species (20%) (FARIA et al., 2014).

The Moraceae family was the most abundant, with 1,072 individuals (34.87 %), and should reduce this population by up to 30 %. Santamour Junior (1990) recommends not exceeding 30% of individuals from the same family to prevent pests and diseases.

Table 2.1. Botanical survey of Joao Pessoa's urban inventory classified by botanical family, species, common name, absolute density (AD), relative density (RD%), absolute frequency (AF), relative frequency (RF%) and phytogeography (E, exotic species, and N, native species).

Family/Species	Common Name	DA	DR (%)	FA	FR (%)	Fitogeo
ANACARDIACEAE		**180**	**5,86**			
Anacardium occidentale L.	cashew tree	20	0,65	18	7,47	N
Mangifera indica L.	hose	75	2,44	28	1,62	E
Schinus terebinthifolias Raddi	beach mastic	79	2,57	53	1,99	N
Spondias mombin L.	cajazeira	3	0,10	3	0,24	N
Spondias purpurea L.	seriguela	1	0,03	1	0,41	E
Spondias tuberosa Rue	umbu	2	0,07	1	0,41	N
ANNONACEAE		**1**	**0,03**			
Annona squamosa L.	pine	1	0,03	1	0,41	E
APOCYNACEAE		**11**	**0,36**			
Himatanthusphagedaenicus (Mart.) Woodson	dairy	3	0,10	3	0,24	N
Plumeria pudica Jacq.	bridal bouquet	5	0,16	1	0,45	E
Plumeria rubra L.	jasmine-mango	2	0,07	2	0,83	E
Thevetia thevetioides (Kunth) K. Schum	Napoleon hat	1	0,03	1	0,41	E
ARAUCARIACEAE		**1**	**0,03**			
Araucaria columnaris (Forest.) Hook.	christmas pine	1	0,03	1	0,41	E
ARECACEAE		**195**	**6,34**			
Acrocomia intumescens Drude	palm. macaiba	46	1,50	2	0,83	N
Caryota sp.	palm. fishtail	2	0,07	1	0,41	E
Cocos nucifera L.	coconut tree	42	1,37	16	6,64	E
Dypsis decaryi (Jum.) Beentje & J. Dransf.	triangular palm	32	1,04	9	3,73	E
Dypsis lutescens (H. Wendl.) Beentje & Dransf.	palm. areca bamboo	8	0,26	2	0,83	E
Pritchardia sp.	palm. fan	11	0,33	3	1,24	E
Roystonea oleracea (Jacq.) O.F. Cook	imperial palm	17	0,55	6	2,49	E
Syagrus romanzoffiana (Cham.) Glassman	palm. jeriva	1	0,03	1	0,41	N
Veitchia merrillii (Becc) H. E. Moore.	Mexican palm	36	1,17	14	5,81	E
BIGNONIACEAE		**362**	**11,78**			
Tecoma stans (L.) Juss. ex Kunth	garden ipe	7	0,23	4	1,66	E
Crescentia cujete L.	gourd	2	0,07	2	0,83	E
Handroanthus chysotrichus (Mart. ex DC.) Mattos	yellow ipe	21	0,68	16	6,64	N
Handroanthus impetiginosus (Mart. ex DC.) Mattos	purple ipe	13	0,42	5	2,07	N
Handroanthus sp.	ipê/ipê-do-serrado	120	3,90	82	34,02	N
Jacaranda sp.	mimosa jacaranda	2	0,07	1	0,41	E
Spathodea campanulata P. Beauv.	african tulip	2	0,07	1	0,41	E
Tabebuia aurea (Silva Manso) Benth. & Hook.f. ex S.Moore	craibeira	107	3,48	6	2,49	N
Tabebuia elliptica (DC.) Sandwith	white ipê	2	0,07	2	0,83	N
Tabebuia pentaphylla (L.) Hemsl.	ipê-de-El-salvador	85	2,77	11	4,56	E
Tabebuia rosealba (Ridl.) Sandwith	periwinkle	1	0,03	1	0,41	N
CANNABACEAE		**1**	**0,03**			
Trema micrantha (L.) Blume	parakeet	1	0,03	1	0,41	N
CAPPARACEAE		**6**	**0,20**			
Crateva trapia L.	cheat	6	0,20	5	2,07	N
CARICACEAE		**1**	**0,03**			
Carica papaya L.	papaya	1	0,03	1	0,41	E
CASUARINAEAE		**28**	**0,91**			
Casuarina sp.	casuarina	28	0,91	7	2,90	E
CHRYSOBALANACEAE		**20**	**0,65**			
Licania tomentosa (Benth.) Fritsch	oyster	20	0,65	13	5,39	N
CLUSIACEAE		**1**	**0,03**			
Clusia sp.	pororoca	1	0,03	1	0,41	N
COMBRETACEAE		**170**	**5,53**			
Terminalia catappa L.	chestnut	170	5,53	117	48,55	E

CUPRESSACEAE		**5**	**0,16**			
Cupressus spp.	cupressus	5	0,16	1	0,41	E

Keep going...

Keep going...

Family/Species	Common Name	DA	DR (%)	FA	FR (%)	Fitogeo
FABACEAE		**580**	**18,87**			
Adenanthera pavonina L.	pigeon eye	40	1,30	31	12,86	E
Albizia lebbeck (L.) Benth	carolina	16	0,52	10	4,15	E
Anadenanthera cf. colubrina (Vell.) Brenan	angico	18	0,59	4	1,66	N
Bauhinia forficata Link	mororo	5	0,16	2	0,83	N
Bauhinia monandra Britt.	cow's foot	10	0,33	6	2,49	E
Bauhinia variegata L	cow's foot	1	0,03	1	0,41	E
Bowdichia virgilioides Kunth	black sucupira	1	0,03	1	0,41	N
Paubrasilia echinata (Lam) E. Gagnin, H. C. Lima & G. P. Lewis	brazilwood	43	1,40	32	13,28	N
Poincianella pluviosa (Benth.) L.P.Queiroz	sibipiruna	9	0,29	9	3,73	N
Caesalpinia pulcherrina (L.) Sw.	flamboyant-mirin	2	0,07	1	0,41	E
Cassia fistula L.	golden cassia	20	0,65	20	8,30	E
Cassia grandis L. f.	cassia-rose	8	0,26	8	3,32	N
Clitoria fairchildiana R.A. Howard	shade tree	20	0,65	18	7,47	N
Delonix regia (Bojer ex Hook.) Raf.	flamboyant	22	0,72	20	8,30	E
Erythrina indica Lam. Var. picta Hort.	Brazilian	25	0,81	6	2,49	E
Hymenaea courbaril L.	jatobà	1	0,03	1	0,41	N
Inga blanchetiana Benth.	ingà	1	0,03	1	0,41	N
Leucaena leucocephala (Lam.) de Wit	leucena	12	0,39	4	1,66	E
Libidibia ferrea (Mart. ex Tul.) L.P.Queiroz	ironwood	3	0,10	2	0,83	N
Pithecellobium dulce (Roxb.) Benth.	hunger killer	15	0,49	10	4,15	E
Prosopis juliflora (Sw.) DC.	algaroba	17	0,55	13	5,39	E
Senna siamea (Lam.) H.S. Irwin & R.C. Barneby	cassia-ferruginea	281	9,14	159	65,98	E
Senna spectabilis (DC.) H.S. Irwin & Barneby	northeast cassia	2	0,07	1	0,41	N
Stryphnodendron sp.	honeycomb	2	0,07	1	0,41	N
Tamarindus indica L.	tamarind tree	6	0,20	5	2,07	E
LAURACEAE		**1**	**0,03**			
Persea americana Mill.	avocado	1	0,03	1	0,41	E
LECYTHIDACEAE		**1**	**0,03**			
Eschweilera ovate (Cambess.) Mart. ex Miers	embiriba	1	0,03	1	0,41	N
LYTHRACEAE		**2**	**0,07**			
Lagerstroemia indica L.	resedà	1	0,03	1	0,41	E
Punica granatum L.	romã	1	0,03	1	0,41	E
MALPIGHIACEAE		**2**	**0,07**			
Byrsonima sp.	myrtle	1	0,03	1	0,41	N
Malpighia emarginata Sessé & Moc. Ex DC..	acerola	1	0,03	1	0,41	E
MALVACEAE		**107**	**3,48**			
Eriothecamacrophylla (K.Schum.) A. Robyns	munguba	1	0,03	1	0,41	N
Talipariti pernambucense (Arruda) Bovini	beach cotton	46	1,50	36	14,94	N
Ceiba sp.	millet tree	1	0,03	1	0,41	N
Guazuma ulmifolia Lam.	mutamba	1	0,03	1	0,41	N
Talipari titiliaceum (L.) Fryxell	beach cotton	19	0,62	12	4,98	E
Luehea ochrophylla Mart.	horse whip horse chestnut	3	0,10	1	0,41	N
Pachira aquàtica Aubl.	maranhão	27	0,88	9	3,73	N
Sterculia sp.	chichà	9	0,29	2	0,83	E
MELIACEAE		**142**	**4,62**			
Azadirachta indica A. Juss	nim	142	4,62	103	42,74	E
MORACEAE		**1072**	**34,87**			
Artocarpus heterophyllus Lam.	jackfruit	3	0,10	3	1,24	E
Ficus benjamina L.	ficus	1054	34,29	196	81,33	E
Ficus elastica Roxb. ex Hornem	ficus elastica	1	0,03	1	0,41	E
Ficus gomelleira Kunth	game tree	12	0,39	12	4,98	E
Morus nigra L.	blackberry	2	0,07	1	0,41	E
MORINGACEAE		**3**	0,10			
Moringa oleifera Lam.	moringa	3	0,10	2	0,83	E
MYRTACEAE		**101**	**3,29**			
Eucalyptus sp.	eucalyptus	3	0,10	1	0,41	E
Psidium guajava L.	guava tree	7	0,23	6	2,49	E
Psidium guineense Sw.	araçà	1	0,03	1	0,41	N
Syzygium cumini (L.) Skeels	olive tree	29	0,94	13	5,39	E
Syzygium jambos (L.) Alston	jambeiro	61	1,98	17	7,05	E

Keep going...

Keep going...

Family/Species	Common Name	DA	DR (%)	FA	FR (%)	Fitogeo
OXALIDACEAE		**1**	**0,03**			
Averrhoa carambola L.	carambola	1	0,03	1	0,41	E
PINACEAE		**1**	**0,03**			
Pinus sp.	pine	1	0,03	1	0,41	E
POLYGONACEAE		**1**	**0,03**			
Triplaris americana L.	ant wood	1	0,03	1	0,41	N
RHAMNACEAE		**2**	**0,07**			
Ziziphus sp.	juazeiro	2	0,07	1	0,41	N
RUBIACEAE		**2**	**0,07**			
Genipa americana L.	genipap	2	0,07	2	0,83	N
SAPINDACEAE		**72**	**2,34**			
Filicium decipiens (Wt. &Arn.) Thwaites	felicia	72	2,34	38	15,77	E
SAPOTACEAE		**1**	**0,03**			
Manilkara zapota (L.) P.Royen	toadstool	1	0,03	1	0,41	N
VERBENACEAE		**1**	**0,03**			
Tectona grandis L. f.	teak	1	0,03	1	0,41	E
Grand Total		**3074**	**100,00**			

Of the species identified in this study, 59 (57.43%) are exotic and 42 (42.57%) are native. Of these, 78.72% are exotic (2,420 specimens) and 21.28% are native (654 specimens). Lindenmaier and Souza (2014) show that, in Cachoeira do Sul/RS, 56.4% are exotic species, which corresponds to 61.7% of the trees; and 43.6% are native species, with 38.3% of the trees. In the survey carried out by Miranda et al. (2015), 76% were exotic and 24% native species in Godoy Moreira/PR. In the city of Patos/PB, 71.4% are exotic species and 28.6% are native (LUCENA et al., 2015). The dominance of exotic species in urban afforestation is mainly due to the use of known and studied species in urban environments, versus the lack of information about native species developing in cities.

According to the bibliographic survey carried out, few Brazilian municipalities have phytogeography different from that found in Joao Pessoa, such as Altonia/PR with 67% native and 33% exotic (MAIORANI et al., 2012); and Nova Esperança/PR, 68% native and 32% exotic (ALBERTIN et al., 2011).

Ficus benjamina was the most abundant species in the city, accounting for 34.29% of the individuals inventoried.

It is the only species outside the technical recommendations described by Grey and Deneke (1978) and Biondi and Althaus (2005), who advise that a species should not exceed 10% to 15% of the total number of individuals in the population.

The abundance of this species was also found in the Santiago neighborhood, in the city of Ji-Paranà/RO, where 39.90% of the individuals were of this species, and in the municipalities of Altamira/PA and Araçoiaba da Serra/SP (PARRY et al., 2012; SILVA et al., 2014; SANTOS JUNIOR; COSTA, 2014).

Other species stand out in the arborization due to their abundance: *Senna siamea, Terminalia catappa* and *Azadirachta indica,* with 9.14% (281 specimens), 5.53% (170 specimens), 4.62% (142 specimens). The four most abundant species account for 53.58% of the trees.

The average linear meter density (LMD) found in the city was 37.66 trees/km of public sidewalk. This value was lower than that found in the Santiago neighborhood, in the city of Ji-Parana/RO, of 120 trees/km (SANTOS JUNIOR; COSTAS, 2014). In the Petrópolis neighborhood, in the city of Natal/RN, it was 70 ind./km (SANTOS *et al.*, 2012) and in the city of Cachoeira do Sul/RS, 53.15 trees/km (LINDENAMAIER; SOUZA, 2014).

However, it was higher than the values found in the Riacho Novo and Bairro Centro neighborhoods, 35 and 26 trees/km, respectively, both located in the city of Nova Iguaçu/RJ (ROCHA et al., 2004). The same happened in the cities of Registro/SP, with 37 ind./km (FERRAZ, 2012) and Senador Guimard/AC, with 6.2 trees/km (MARANHO et al., 2012).

These values are an index used to measure the density of trees on urban sidewalks, where, according to (LINDENAMAIER; SOUZA, 2014), there is no ideal or recommended number for this index. However, according to Paiva (2009), the Brazilian Society of Urban Afforestation (SBAU) recognizes at least 100 trees per km of sidewalk as the ideal number. In comparison with other locations, it can be said that the city of Joao Pessoa has a medium density and is still well below the number considered ideal by the SBAU.

In the population analyzed, 475 individuals were identified as having phytosanitary problems caused by insects such as termites, mealybugs, leaf-cutting ants or bedbugs, corresponding to 15.45% of the population. 19.39% of the specimens showed apparent problems caused by fungi or other microorganisms on the leaves, trunks and neck of the tree. 71.50% (2,198 specimens) of the population were free of insects and microorganisms and 28.50% had some kind of pathogen. In the central area of the city of Sao Joaquim/SC, 78% of the individuals observed showed some kind of pest and disease attack (SOUZA et al., 2014), a figure higher than that found in Joao Pessoa/PB.

F. benjamina was the species with the highest number of individuals with phytopathological problems: 13.66% of the species' population had problems caused by insects, and 15.65% had problems caused by microorganisms.

The *S. siamea* and *T. pentaphylla* species come next in the number of individuals with insect problems, 44 (15.66%) and 43 (50.59%) specimens, and the *Handroanthus* spp. and *S. siamea* species come next in the number of individuals with microorganism problems, with 53 (44.17%) and 48 (17.08%) individuals, respectively.

With regard to conflicts with urban equipment, 1,789 individuals (58.10% of the population) had problems, with the most common being conflicts with the electricity grid, lifting or breaking sidewalks, water lines and sewage systems. With regard to physical damage, 27.94% of the population had problems with some structure, the most common being objects stuck in trunks and mechanical impacts, corresponding to 859 specimens. Around 36.10% (1,110 individuals) of the population had none of these problems.

In the Santiago neighborhood, in the municipality of Ji-Paranâ/RO, 30% of the population had conflicts with overhead telecommunications and electricity cables (SANTOS JUNIOR; COSTA, 2014). In the city of Nova Iguaçu/RJ, 35% of the trees had conflicts with aerial infrastructure (ROCHA *et al.,* 2004). In the central area of the municipality of Sao Joaquim/SC, 41% of the specimens that make up the urban arborization showed some kind of injury caused by vandalism (SOUZA et al., 2014). Nascimento *et al.* (2014) found that in the Central District of the city of Resende/RJ, 55% of the population showed signs of injury or mechanical aggression.

The individuals with the greatest urban conflicts or physical damage were *F. benjamina, S.siamea* and *T.catappa,* with 628, 213 and 114 specimens, respectively, in conflicts with urban equipment, and 244, 73 and 61 individuals with physical damage, respectively.

The study also indicated that 3.51% of the population had problems caused by insects and microorganisms, urban equipment and physical damage, corresponding to 108 individuals, and 842 had no harmful problems, corresponding to 27.39% of the population.

The *F. benjamina* species had the highest number of individuals in conflict with the city's infrastructure. This proves that this species is not adapted to planting trees on sidewalks and central medians. In the cities of Cafeara/PR and Tuparendi/RS, and in the neighborhood of Ferraropolis, in the municipality of Garça/SP, the species in question is also responsible for most of the conflicts with overhead power lines and damage to sidewalks (LOCASTRO et al., 2014; NUNES et al., 2013; MOTTER; MULLER,2012).

Some authors report that *F. benjamina* is an unsuitable species for tree planting in streets and avenues, due to its aggressive and superficial root system and foliage that can cause gutters, culverts and manholes to clog. This includes species of the *Ficus* sp. genus, which have a strong tendency to cause future problems (SARTORI; BALDERI, 2011; LORENZI et. al., 2003; SANTOS;TEIXEIRA, 2001).

The Shannon diversity index H' found in the city's trees was 2.96 nats/ind. and the Pielou Equability J' was 0.37. In comparison with other cities, the municipality can be considered to have an average diversity index. In Cachoeira do Sul/RS, the estimated value of H' was 3.14 nats/ind. (LINDENMAIR; SOUZA, 2014), and in Colorado/RS, the diversity index was 2.95 nats/ind. (RABER; REBELATO, 2010). In the city of Sorocaba/SP (CARDOSO-LEITE *et al.,* 2014), the H' index found was 3.73 nats/ind., which is considered a high index for the standard found in Brazilian municipalities.

The results by sector (Table 2.2) indicate that sector A had the lowest linear meter density (LMD), 32.50 trees/km, and sector B had the highest linear density, 46.80

trees/km. The CeD sectors had a linear meter density of 43.20 and 36.20 trees/km, respectively. This indicates an uneven distribution of urban trees in the city, with greater concentration in some neighborhoods and sparseness in others.

This difference between the results can be explained by the use and occupation of the land in each sector, being directly influenced by population density in some neighborhoods, existing urban infrastructure such as paved streets and basic sanitation, and more frequent use such as commercial, industrial or residential areas.

With regard to physical health aspects, sector D had the highest relative density of individuals with problems caused by insects, and sector C had the lowest relative density. Sector A had the highest rate of individuals suffering from apparent microorganism attacks, and sector D the lowest.

Sectors A and D had the highest relative numbers of individuals in conflict with urban facilities, and sector C had the lowest. Sector A had the highest rate of individuals with physical injuries, and sector D had the lowest.

Table 2.2. Number of individuals identified as having tree-planting problems in the city of Joao Pessoa/PB by sector analyzed.

Sectors	N.I.	N.I. (%)	N.P.	P.N. (%)	N.C.	N.C. (%)	N.D.	N.D. (%)	Km Traveled	DML
A	164	13,39	328	26,78	966	78,86	420	34,29	37,70	32,50
B	112	15,41	142	19,53	218	29,99	182	25,03	15,50	46,80
C	73	12,59	86	14,83	187	32,24	137	23,62	13,40	43,20
D	126	23,25	40	7,38	415	76,57	120	22,14	14,95	36,20
Total	475	15,45	596	19,39	1786	58,10	859	27,94	81,60	37,66

Where: Absolute number of individuals with insects (N.I.) and relative [N.I. (%)]; Absolute number of individuals with pathogens (N.P.) and relative [N.P. (%)]; Absolute number of individuals in conflicts with urban equipment (N.C.) and relative [N.C. (%)]; Absolute number of individuals with physical damage (N.D.) and relative [N.D. (%)]; Km traveled in each sector (km); and Density linear meter found in each sector (DML).

All these results together indicate that the afforestation of the neighborhoods located in sector A needs greater attention from the municipal authorities. This is due to the high number of individuals in conflict with urban equipment, physical damage and pathology caused by microorganisms. The greatest danger in this area is the toppling of trees during rainy periods and strong winds, which can cause material damage and put people's lives at risk.

Individuals of the species *F. benjamina, S. siamea, T. pentaphylla, Handroanthu* spp. and *T. catappa* deserve special attention for having the highest absolute density, individuals with phytopathological problems, conflicts with urban equipment and physical damage. Periodic maintenance and replacement actions are being carried out in all regions of the municipality.

In the municipality of Goiânia/GO, the Urban Forestry Master Plan recommends the immediate removal of individuals of the species *F. benjamina* and *T. catappa,* because they have been planted in compacted soil, with rubble and construction waste, generating a large amount of waste, among other things (GOIÂNIA, 2008).

The maintenance of urban tree plantations will result in the generation of a large amount of biomass waste, which will increase costs for the municipality. Studies show the feasibility of using this material to generate electricity, organic compost, biogas, briquettes and alternative fuel (ARAÙJO et al., 2013; CORTEZ, 2011; TORRÊS FILHO, 2005), which is an alternative to avoid increasing public spending.

2.3.2João Pessoa's urban forestry improvement plan

Due to the heterogeneous distribution of urban trees in the municipality and in order to obtain more precise information, it would be advisable to carry out a tree census. Even so, it was possible to identify the species that presented the greatest problems for public and public health facilities, and the species that presented the least problems. Table 2.3. shows the list of the 30 most representative species in terms of individuals, with health problems, physical damage and conflicts with urban facilities. This list gives us an indication of which species are more adapted to the municipality, as they present fewer problems, and which are less adapted, with more urban conflicts and phytosanitary problems.

The 30 most representative species in terms of number of individuals accounted for

94% of the sample. The percentage of individuals of each species was taken into account in order to identify those that are most adaptable and resistant to pathogens in the city. The table shows that the species *A. intumescens*, A. *colubrina,* Handroanthus sp., *T. pentaphylla, A. lebbeck, P. aquàtica, T. aurea, A. pavonina, D. regia, C. fairchildiana, J. juliflora, P. dulce, E. indica, S. maloccense* and *S. terebinthifolia had* the highest percentages of individuals with pathogenic problems (microorganisms) and insect attacks. This indicates that they are the species most susceptible to attack by microorganisms in the municipality.

The species *R. oleraceae, M. indica, S. siamea, A. indica, P juliflora, A. pavonina, L. tomentosa, F. decipiens, S. molaccense, T. catappa, P. dulce, E. indica, D. regia, F. bejamina, A. colubrina* and *D. regia* showed the highest percentages of individuals in conflict with public equipment and physical damage. This result could be due to the species' lack of adaptability to the urban environment or a lack of planning when planting trees.

The census could be carried out in parts, covering 20 to 25% of the city's streets each year, so the study would be completed in 4 to 5 years. As the results are analyzed, with the identification and location of individuals in need of action (corrective or preventive) or replacement, the tree maintenance team could be activated. By carrying out tree maintenance after the census, the volume of wood will increase in the first 5 years. However, there would be a reduction over time because there would no longer be an excessive number of individuals in need of maintenance, which would reduce costs and the amount of pruning and replacement.

Table 2.3. List of the 30 most representative species with phytosanitary problems, conflicts with urban equipment and physical damage (Ind.).

Scientific name	Common name	N	N.I.	N.I.%	N.P.	N.P.%	N.C.	N.C.%	N.D.	N.D.%
F. benjamina	ficus	1054	144	13,7	165	15,7	628	59,58	244	23,15
Handroanthus sp.	ipê	120	25	20,8	53	44,2	69	57,5	50	41,67
S. siamea	cassia ferruginha	281	44	15,7	48	17,1	213	75,8	73	25,98
A. intumescens	palm. Macaiba	46	4	8,7	38	82,6	11	23,91	29	63,04
T. aurea	craibeira	107	5	4,67	37	34,6	15	14,02	60	56,07
T. pentaphylla	ipê-de-El- salvador	85	43	50,6	37	43,5	49	57,65	48	56,47

T. catappa	chestnut	170	27	15,9	30	17,6	114	67,06	61	35,88
A. indica	nim	142	12	8,45	18	12,7	103	72,54	31	21,83
M. indica	hose	75	13	17,3	15	20	58	77,33	22	29,33
A. pavonina	pigeon eye	40	7	17,5	12	30	28	70	18	45,00
S. malaccense	jambeiro	61	16	26,2	11	18	42	68,85	24	39,34
F. decipiens	felicia	72	11	15,3	10	13,9	50	69,44	11	15,28
P. aquatica	castanheira-do-maranhao	27	2	7,41	10	37	9	33,33	7	25,93
A. colubrina	angico	18	3	16,7	10	55,6	7	38,89	13	72,22
S. terebinthifolia	beach mastic	79	18	22,8	8	10,1	37	46,84	18	22,78
E. indicates	Brazilian	25	7	28	6	24	16	64	6	24,00
D. regia	flamboyant	22	5	22,7	6	27,3	14	63,64	15	68,18
A. lebbeck	carolina	16	4	25	6	37,5	5	31,25	6	37,50
V. merrillii	palm. Mexican	36	1	2,78	5	13,9	8	22,22	1	2,78
S. jambolanium	olive tree	29	4	13,8	5	17,2	9	31,03	2	6,90
C. fairchildiana	shade tree	20	5	25	5	25	11	55	12	60,00
C. equisetifolia	flamboyant- mirim	28	1	3,57	4	14,3	16	57,14	8	28,57
A. occidentale	cashew tree	20	2	10	4	20	10	50	8	40,00
T. tiliacium	beach cotton	19	3	15,8	4	21,1	11	57,89	2	10,53
P. juliflora	algaroba	17	8	47,1	4	23,5	12	70,59	5	29,41
T. Pernambuco	beach cotton	46	5	10,9	3	6,52	24	52,17	13	28,26
P. echinata	brazilwood	43	1	2,33	3	6,98	17	39,53	6	13,95
C. nucifera	coconut tree	42	2	4,76	3	7,14	23	54,76	3	7,14
C. fistula	golden cassia	20	4	20	3	15	10	50	2	10,00
H. chysotrichus	yellow ipe	21	4	19	2	9,52	13	61,9	2	9,52
P. dulce	hunger killer	15	5	33,3	2	13,3	10	66,67	1	6,67
L. tomentosa	iotizer	20	3	15	1	5	14	70	13	65,00
D. decaryi	palm. Triangular	32	0	0	0	0	9	28,13	0	0,00
R. oleracea	palm. Imperial	17	0	0	0	0	14	82,35	0	0,00
H. impetiginosus	purple ipe	13	1	7,69	0	0	5	38,46	1	7,69

Where: Number of individuals (N); Absolute number of individuals with insects (N.I.) and relative [N.I. (%)]; Absolute number of individuals with pathogens (N.P.) and relative [N.P. (%)]; Number of individuals with insects (N.I.) and relative [N.I. (%)]; Number of individuals with pathogens (N.P.) and relative [N.P. (%)].

absolute number of individuals in conflicts with urban equipment (N.C.) and relative numbers [N.C. (%)], and; absolute number of individuals with physical damage (N.D.) and relative numbers [N.D. (%)].

Based on the findings of the study, the following actions are recommended as a management plan proposal:

1. Replacing the individuals of the least suitable species with the most suitable species, as listed in Table 3.3;

2. Carrying out a partial census every year, 20 to 25%/year, and implementing the maintenance of individuals identified with problems, after identification;

3. Development of an arboricultural monitoring program, by registering the individuals catalogued in the arboricultural census;

4. Development of a program to raise environmental awareness of urban afforestation, with the indication of species adapted to urban areas;

5. Workforce qualification and training program for maintenance services, following the guidelines described in ABNT NBR 162461:2013;

6. Partnership between the public and private sectors for the implementation of afforestation and maintenance projects, and;

7. Development of a cooperative to add value to urban forestry waste.

2.4 CONCLUSIONS

From the analysis carried out, it can be concluded that:

The urban forest is unevenly distributed throughout the municipality, concentrated in some neighborhoods and deficient in others;

Its floristic richness is considered average compared to other localities, with *F. benjamina*, *S. siamea* and *T. catappa* standing out in terms of abundance;

H' diversity index, considered to be average, DML of 37.66 ind./km and the species *F. benjamina*, *S. siamea*, *T. pentaphylla* and *T. catappa* showed the highest indices with phytopathological problems, physical damage and conflicts with urban facilities;

The proposed urban tree management plan took into account the environmental functionality of tree planting, the dynamics of the city and sustainability in urban areas. There should be more discourse (technical and economic) on its implementation and execution.

2.5 REFERENCES

ANGIOSPERM PHYLOGENY GROUP (A.P.G) IV. 2016. An update of the Angiosperm Phylogeny Group classification for the orders and families of flowering plants: APG IV.

Botanical Journal of the Linnean Society, London, v. 181, p. 1-20, 2016.

ALBERTIN, R. M.; De ANGELIS F.; De ANGELIS NETO R.; De ANGELIS B. L. D. Quali-quantitative Diagnosis of Street Tree Planting in Nova Esperança, Paraná, Brazil.

Revista da Sociedade Brasileira de Arborizaçao Urbana, v.6 n. 3, p. 128-148, 2011.

ANGELIS, B. L. D.; SAMPAIO, A. C. F.; TUDINI, O. G.; ASSUNÇÂO, M. G. T.; ANGELIS NETO, G. de. Evaluation of trees on public roads in the central area of Maringa, state of Paraná: estimate of waste production and final destination. **Acta. Sci. Agron**, v. 29, n. 1, 2007.

ARAÙJO, V. C.; BEZERRA, E. S. P.; LIMA NETO, J. A.; ILARIANO, R. N. S.; VASCONCELO, Z. N. F.; VALE, M. B. Study of the use of tree prunings to produce briquettes in two municipalities in Rio Grande do Norte. In: IX IFRN Scientific Initiation Congress. **Anais ...**Currais Novos/RN, 0673-0679, 2013.

BARROS, V. C. C. **Briquettes produced from urban pruning waste and cardboard packaging**. 2013. Monograph (Forestry Engineering). University of Viçosa, Viçosa, 2013.

BATISTEL, L. M.; DIAS, M. A. B.; MARTINS, A. S.; RESENDE, I. S. M. Qualitative and quantitative diagnosis of urban afforestation in the Promissao and Pedro Cardoso neighborhoods, Quirinòpolis, Goias. **Revista da Sociedade Brasileira de Arborizaçao Urbana**, v.4, n.3, 110-129, 2009.

BENATTI, D. P. B.; TONELLO, K. C.; ADRIANO JUNIOR, F. C.; SILVA, J. M. S.; OLIVEIRA, I. R.; ROLIM, E. N.; FERRAZ, D. L. Urban tree inventory in the municipality of Salto de Pirapora, SP. **Revista Arvore**, v. 36, n. 5, p.887-894, 2012.

BOBROWSKI, R.; BIONDI, D. Characterization of the planting pattern adopted in street tree planting in Curitiba, Paranâ. **Revista da Sociedade Brasileira de Arborização Urbana**, v.7, n.3, 2012.

BOBROWSKI, R. **Structure and dynamics of street tree planting in Curitiba, Paranâ, in the period 1984-2010**. Curitiba, 2010. Dissertation (Master's degree in Forestry Engineering).

Federal University of Paranâ, Curitiba, 2010.

BIONDI, D. **Diagnosis of street tree planting in the city of Recife**. Dissertation (Master's Degree in Forestry Sciences). Federal University of Paranâ, Paranâ, 1985.

BIONDI, D.; ALTHAUS, M. **Curitiba's street trees**: cultivation and management. Curitiba: FUPEF, 2005. 177p.

CARDOSO-LEITE, E.; FARIA, L. C.; CAPELO, F. F. M.; TONELO, K. C.; CASTELLO, A. C. D.; Floristic composition of urban trees in Sorocaba/SP, Brazil. **Revista da Sociedade Brasileira de Arborizaçâo Urbana**, v. 9, n. 1, 2014.

MINAS GERAIS ENERGY COMPANY. **Afforestation manual.** Belo Horizonte: CEMIG/Fundaçao Biodiversitas, 2011.

CORTEZ, C. L. **Study of the potential use of biomass resulting from tree pruning for energy generation:** case study: AES Eletropaulo. 2011. Thesis (Doctorate in Science). University of Sao Paulo, Sao Paulo, 2011.

DANTAS, I. C.; SOUZA, C. M. C.. Urban afforestation in the city of Campina Grande, PB: Inventory and species. **Revista de Biologia e Ciência da Terra**, v. 4, n. 2, 2004.

FARIA, R. F.; SOUSA, V. R.; MIRANDA, S. C. Urban afforestation in the city of Itapuranga, Goiàs. **Revista da Sociedade Brasileira de Arborizaçâo Urbana**, v. 9, n. 2, 2014.

FERRAZ. M. C. Inventory of urban trees in the city of Registro-SP. **Revista da Sociedade Brasileira de Arborização Urbana**, v. 7, n. 2, 2012.

GOIÂNIA. Goiânia City Hall. **Urban Forestry Master Plan for Goiânia**. Goiânia, 2008. Available at: <http://www.goiania.go.gov.br/download/amma/relatorio_Plano_Diretor.pdf>. Accessed on: January 5, 2017.

GREY, G. W.; DENEKE, F. J. **Urban forestry**. New York: John Wiley, 1978.

HUECK, K. **The Forests of South America:** Ecology, Composition and Economic Importance. Sao Paulo: University of Brasilia; Poligono, 465p, 1972.

BRAZILIAN Institute of Geography AND STATISTICS. **Municipal population census:** Joao Pessoa. 2016. Available at: <http://cidades.ibge.gov.br/xtras/perfil.php?codmun=250750>. Accessed on: April 5, 2017.

JOÂO PESSOA. Municipal Planning Department. **Cadastral file of the urban streets of Joâo Pessoa/PB.** 2015. Delivered in person.

JOÂO PESSOA. Municipal Department of Communication. **City Hall exceeds target and carries out 9.2 thousand tree prunings in 2012**. SECOM, 2012. Available at: <www.joaopessoa.pb.gov.br>. Accessed on: January 5, 2017.

LIMA NETO, E. M.; BIONDI, D.; LEAL, L.; SILVA, F. L. R.; PINHEIRO, F. A. P. Análisis da composiçao floristica de Boa Vista-RR: subsidio para a gestão da arborizaçao de ruas. **Revista da Sociedade Brasileira de Arborização Urbana**, v. 11, n. 1, p. 58-72.

LINDENMAIER, D. S.; SANTOS, N. O. Urban tree planting in the squares of Cachoeira do Sul-RS-Brazil: Phytogeography, Diversity and Green Areas Index. **Botanical Research** No. 59, São Leopoldo: Instituto Anchietano de Pesquisas, 2008.

LINDENMAIER, D. S.; SOUZA, B. S. P.; Road Tree Planting in Cachoeira do Sul/RS: Diversity, Phytogeography and Conflicts with Urban Infrastructure. **Revista da Sociedade Brasileira de Arborização Urbana**, v. 9, n. 1, 2014.

LOCASTRO, J. K.; RASBOLD, G. G.; PERREIRA, J. S. R.; SOARES, B.; CAXAMBÙ, M. G. Census of urban afforestation in the municipality of Cafeara, Paranâ. **Revista da Sociedade Brasileira de Arborização Urbana,** v. 9, n. 3, 2014.

LORENZI, H.; SOUZA, H. M.; TORRES, M. A. V.; BACHER, L. B.; **Arvores exóticas no BRASIL:** Madeiras, ornamentais e aromàtica. Nova Odessa, SP: Instituto Plantarum, 2003.

LORENZI, H.; NOBLICK, L. R.; KAHN, F. K.; FERREIRA, E.; **Flora Brasileira Lorenzi: Arecaceae (palm)**. Nova Odessa, SP: Instituto Plantarum, 2010.

LUCENA, J. N.; SOUTO, P. C.; CAMANO, J. D. Z.; SOUTO, J. S.; SOUTO, L. S. Afforestation in central flowerbeds in the city of Patos, Paraiba. **Green Journal of Agroecology and Sustainable Development**, v. 10, n. 4, 2015.

MAIORANI, E.; WESOLOWSKI, J.B.; MELISINAS, V.A.P.dos S.; FABRIN, T.C.; GASQUES, L.S. Survey of urban afforestation in the municipality of Altônia-PR.

Publicatio UEPG Ciências Biológicas e da Saùde, Ponta Grossa-PR, v.18, n. 2, p. 101108, 2012.

MARANHO, A. S; PAULA, S. R. P.; LIMA, E.; PAIVA, A. V.; ALVES, A. P.; NASCIMENTO, D. O. Census survey of urban road tree planting in Senador Guimard, Acre. **Revista da Sociedade Brasileira de Arborização Urbana**, v. 7, n. 3, 2012.

MIRANDA, Y. C.; MACHADO, M. S.; SAILVA, L. S.; ESTEVAM, R.; MARTINS NETO, F. F.; CAXAMBU, M. G. Qualitative and quantitative analysis of street tree planting in the municipality of Godoy Moreira - PR. **Revista da Sociedade Brasileira de Arborização Urbana**, v. 10, n. 1, 2015.

MORAIS, L. A.; MACHADO, R. R. B. Urban afforestation in the municipality of Timon/MA: inventory, diversity and quali-quantitative diagnosis. **Revista da Sociedade Brasileira de Arborização Urbana**, v. 9, n. 4, 2014.

MOTTER, N.; MÜLLER, N. G. Diagnosis of urban afforestation in the municipality of Tuparendi-RS. **Revista da Sociedade Brasileira de Arborização Urbana**, v.7, n.4, 2012.

NASCIMENTO, M. S.; RODRIGUES, E. R.; SOUZA, C. A.; FARIAS, M. J. B.; PEDERASSI, J.; LIMA, M. S. C. Análisis quali-quantitativa da arborização das áreas pùblicas do Bairro Centro de Resende, RJ. **Revista da Sociedade Brasileira de Arborizaçâo Urbana**, v. 9, n. 4, 2014.

NOWAK, D. J.; WALTON, J. T.; STEVENS, J. C.; CRANE, D. E.; HOEHN, R. E. Effect of plot and sample size on timing and precision of Urban Forest Assessments.

Arboriculture & Urban Forestry, v. 34, n. 06, p. 386-390, 2008.

NUNES, R. L.; MARMONTEL, C. V. F.; RODRIGUES, J. P.; MELO, A. G. C.

Qualitative and quantitative survey of the urban afforestation of the Ferraropólis neighborhood in the city of Graça-SP. **Revista da Sociedade Brasileira de Arborizaçâo Urbana**, v.8, n.1, 2013.

OLIVEIRA FILHO, P. C.; SILVA, S. V. K. An information system for spatial and decision-making support for urban tree management in the municipality of Guarapuava, Paranâ.

Revista da Sociedade Brasileira de Arborizaçâo Urbana, v. 5, n. 3, p. 82-96, 2010.

PAIVA, A. V. Aspectos da arborizaçao urbana do Centro de Cosmópolis - SP. **Revista da Sociedade Brasileira de Arborizaçâo Urbana**, v. 4, n. 4, p. 17-31, 2009.

PARRY, M. M.; SILVA, M. M.; SENA, I. S.; OLIVEIRA, P. M. Composiçao floristica da arborizaçao da cidade de Altamira, Parà. **Revista da Sociedade Brasileira de Arborização Urbana**, v. 7, n. 1, 2012.

PERIOTTO, F; MESTRINER, M. M.; HELMANN, A. C.; SANTOS, T. O.;

BORTOLOTTI, S. L. Analysis of urban afforestation in the municipality of Medianeira, Paranâ.

Revista da Sociedade Brasileira de Arborizaçâo Urbana, v. 11, n. 2, 59-74, 2016.

QUISSINDO, I. A. B.; OCONOR, E. F.; LUNA, D. P. Evaluation of tree vegetation in the main streets of the city of Huambo-Angola. **Revista da Sociedade Brasileira de Arborização Urbana**, v. 11, n. 1, p. 43-57, 2016.

RABER, A.P.; REBELATO, G.S. Arborizaçao viària do municipio de Colorado, RS - BRASIL: análisis quali-quantitativa. **Revista da Sociedade Brasileira de Arborização Urbana**, v.5, n.1, 2010.

RECIFE. Secretariat for the Environment and Sustainability (SEMAS). **Manual de arborização:** orientação e procedimentos técnicos para a implantação da cidade do Recife. Secretariat for the Environment and Sustainability (SEMAS) Recife: [s.n.], 2013.

REFLORA. **List of Species of the Flora of Brazil.** Rio de Janeiro Botanical Garden Research Institute. 2017. Access at:<http://reflora.jbrj.gov.br/jabot/listaBrasil/ConsultaPublicaUC/ConsultaPublicaUC.do> . Accessed on: February 10, 2017.

ROCHA, A. T.; SANTOS, P. S.; NETO, S. N. O. Arborizaçao de vias publica em Nova Iguaçu, RJ: o caso dos Bairros Rancho Novo e Centro. **Revista Arvore**, v. 28, n. 4, p. 599607, 2004.

ROSSETTI, A. I. N.; PELLEGRINO, P. R. M.; TAVARES, A. R. Trees and their interfaces in the urban environment. **Revista da Sociedade Brasileira de Arborização Urbana**, v. 5, n. 1, 2010.

SANTAMOUR JUNIOR, F.S. Trees for urban planting: diversity uniformity, and common sense. **Proceedings...** In: METRIACONFERENCE, 7., 1990, Lisle. Proceedings. Lisle: p.57-66. 1990.

SANTOS, T. O. B.; LISBOA, C. M. C. A.; CARVALHO, F. G. Analysis of the road arborization of the Petrópolis neighborhood, Natal, RN: An approach for diagnosis and planning of urban flora. **Revista da Sociedade Brasileira de Arborização Urbana**, v. 7, n. 4, 2012.

SANTOS, N. R. Z; TEIXEIRA, I. F. **Arborização de Vias Pùblicas:** Environment X Vegetation, Porto Alegre: Editora Pallotti, 2001.

SANTOS, J. S.; SANTOS, G.D. Estudo microclimàtico em Pontos Representativos da malha urbana da cidade de Joao Pessoa\PB: Uma avaliaçao do campo térmico. **Revista Brasileira de Geografia Fisica**, v. 6, n. 5, p. 1430-1448, 2013.

SANTOS JUNIOR, A.; COSTA, L. M.; Species used in the urban afforestation of Bairro Santiago, Ji-Paranâ/RO. **Revista da Sociedade Brasileira de Arborizaçâo Urbana**, v. 9, n. 1, 2014.

SARTORI, R. A.; BALDERI, A. P. Inventory of urban afforestation in the municipality of Socorro-SP and proposal for an index of damage to city infrastructure. **Revista da Sociedade Brasileira de Arborização Urbana**, v.6, n.4, 2011.

SALVI, L.T.; HARDT, L.P.A.; ROVEDDER, C.E.; FONTANA, C.S. Tree planting along streets - Túneis Verdes- in Porto Alegre, RS, Brazil: quantitative and qualitative evaluation. **Revista Arvore**, v.35, n.2, p.233-243. 2011.

SÂO PAULO (Capital). Municipal Department of Greenery and the Environment. **Technical Manual for Urban Afforestation**. Sâo Paulo, 2015. Available at: <http://www.prefeitura.sp.gov.br/cidade/secretarias/meio_ambiente/publicacoes_svma/ind ex.php?p=188452>. Accessed on: January 07, 2017.

SCHALLENBERGER, L. S.; ARAÙJO, A. J; ARAÙJO, M. N.; DAINER. L. J.;

MACHADO, G. O. Evaluation of the condition of urban trees in the main parks and squares in the municipality of Irati/PR. **Revista da Sociedade Brasileira de Arborização Urbana**, v.5, n.2, 105-123, 2010.

SILVA, T. G.; LEITA, E. C.; TONELLO, K. C. Inventory of urban afforestation in the municipality of Araçoiaba da Serra, SP. **Revista da Sociedade Brasileira de Arborização Urbana**, v. 9, n. 4, 2014.

SILVA FILHO, D. F.; PIZETTA, P. U. C.; ALMEIDA, J. B. S. A.; PIVETTA, K. F. L.; FERRAOUDO, A. S. Related database for registration, evaluation and management of trees on public roads. **Revista Arvore**, v. 26, n. 5, p. 629-642, 2002.

SOUZA, A. M.; NACHTERGAELE, M.F.; CARBONI, M. Inventàrio Florestal da arborçâo do municipio de Jaù/SP. Jaù: Instituto Pró-Terra & Secretaria do Meio Ambiente - SEMEIA. **Technical report**, 2004.

SOUZA, P. F.; BOURSCHEID, C. B.; POMPÉO, P. N.; STANG, M. B.; MANFROI, J.; RODRIGUES, M. D. S.; SILVA, A. C.; HIGUCHI, P. Inventory and Recommendations for the Afforestation of the City Center of Sâo Joaquim, SC. **Revista da Sociedade Brasileira de Arborização Urbana**, v. 9, n. 4, 2014.

SOUZA, J. F.; SILVA, R. M.; SILVA, A. M. Influence of land use and occupation on surface temperature: the case study of Joâo Pessoa-PB. **Ambiente Construido**, v. 16, n. 1, p. 21-37, 2016.

TÔRRES FILHO, A. **Technical and environmental feasibility of using wood waste to produce an alternative fuel**. 2007, 61 f. Dissertation (Master's Degree in Sanitation, Environment and Water Resources). Federal University of Minas Gerais, Belo Horizonte, 2005.

TROPICOS. **Missouri Botanical Garden**. 2016. Available at: <http://tropicos.org>. Accessed on: March 5, 2017.

TOSCAN, M. A. G.; RICKLI, H. C.; BARTINICK, D. SANTOS, D. S.; ROSSA, D. Inventory and analysis of the afforestation of the Vila Yolanda neighborhood, in the municipality of Foz do Iguaçu-PR. **Revista da Sociedade Brasileira de Arborização Urbana**, v. 5, n. 3, p. 165184, 2010.

3. ARTICLE 2 - CARBON FATIGUE FOR FOUR SCENARIOS OF URBAN FLOODING WASTE IN JOÂO PESSOA

YURI ROMMEL VIEIRA ARAÙJO; MONIJANY LINS GÓIS; LUIZ MOREIRA COELHO JUNIOR; MONICA CARVALHO

SUMMARY

The improper disposal of urban tree waste causes significant environmental impacts during its decomposition. The aim of this study was to quantify the environmental impacts and analyze four scenarios for urban tree waste in Joao Pessoa. The IPCC 2013 GWP 100a used in the life cycle assessment was used to quantify the environmental impacts of the disposal scenarios for urban tree maintenance waste. The end-of-life treatments (final disposal scenarios) were: landfill without and with methane collection, simple municipal incineration and wood reuse. The results indicate that: simple landfill generated 3,690 t CO_2 -eq (136.34 kg CO_2/t collected), causing the greatest impact; landfill with methane collection released 3.070 t $_{CO2-eq}$ (113.43 kg $_{CO2/t}$ collected); for municipal incineration, the result indicated emissions of 1,930 t CO_2 -eq (71.31 kg CO_2 /t); and reuse is the process with the least environmental impact, emitting 753 t CO_2 -eq (27.82 kg CO_2/t collected). The best disposal options, with the lowest greenhouse gas emissions, were the reuse of urban tree waste, with the lowest environmental impact, and the incineration scenario. The study showed that the reuse of biomass, as well as being environmentally viable, has the potential to contribute to the city's environmental quality, including the possibility of being used as carbon credits.

Key words: Biomass, life cycle assessment, greenhouse gases

3.1 INTRODUCTION

Waste from domestic activities and urban cleaning is technically known as municipal solid waste (MSW) (ABRELPE, 2015). Most Brazilian municipalities find it difficult to manage MSW, due to the complexity and scope of the issue. Proper management of urban tree waste is not always a priority on municipal agendas (MEIRA, 2010).

Urban afforestation involves any form of vegetation located in urban open spaces, being constructive elements of the landscape (SÂO PAULO, 2015). The maintenance of these areas requires periodic actions such as irrigation, complementary fertilization, pruning and replacement of individuals (COPEL, 2009).

The purpose of removing branches, fruit, inflorescences or foliage is to promote the longevity of the tree and guarantee environmental services in the urban area. The waste generated from this activity accounts for a significant portion of MSW and is disposed of in landfills and garbage dumps, and in some cases is incinerated (BARATTA JUNIOR; MAGALHÂES, 2010; CEMIG, 2011; RECIFE, 2013).

In 2014, urban tree waste in the municipality of Goiana (PE) accounted for 1.41% of the MSW sent to processing plants. In Jaboatao dos Guararapes (PE), 1.48 % of urban solid waste was generated by urban trees (SNSA, 2016). The disposal of tree waste in landfills or other disposal sites represents a high cost for Brazilian municipalities. In addition to being a waste of material with energy potential to be used in industry or households (MARTINS, 2013).

In 2010, the average cost of disposing of urban forestry waste in landfills was R$68.00/t in the state of São Paulo (MEIRA, 2010). In Joao Pessoa (PB), this cost corresponds to R$ 200.00/t for the Joao Pessoa metropolitan landfill (EMLUR, 2016).

From 2008 to 2014, the municipality of Joao Pessoa (PB) generated an accumulated total of 127x103 t of urban tree waste. According to Joao Pessoa's Municipal Plan for Integrated Solid Waste Management, urban tree waste should be sent to a sorting center by 2022, with EMLUR responsible for its proper disposal (EMLUR, 2014).

MSW has the potential to cause environmental impacts and damage to health when disposed of improperly or misused, and can contribute to climate change. The environmental impact occurs during the decomposition of MSW, with the potential to contaminate the soil and surface and groundwater, the formation of toxic, asphyxiating and explosive gases, and the generation of greenhouse gases (GHG), mainly methane (CH_4) (GOLVEIA, 2012). These environmental impacts can be quantified by developing a life cycle assessment (LCA) of the waste.

Life Cycle Assessment (LCA) is an important methodology for measuring the environmental impacts of activities and processes. It is based on the idea that improvements in a given process can induce secondary effects throughout the life cycle in a positive and/or negative way, affecting the environmental performance of goods and services (CARVALHO *et al.,* 2015*;* FREIRE *et al.,* 2015; PIRES *et al.,* 2002). The life cycle can be defined as the stages necessary for a product or service to be developed or designed, fulfill its respective function and reach the stage of disposal, recycling or reuse.

LCA is an appropriate methodology for assessing environmental impacts from the extraction of natural resources to the final disposal of the product, including all stages of the production system (GUINÉE, 2001; GUINÉE, 2002). This method has applications in product development and improvement; defining strategic planning and public policies; managing the environmental impacts of products and services and responsible ecological marketing (ACV BRASIL, 2016; CARVALHO *et al.,* 2015; FREIRE et *al*., 2015).

According to the scientific literature on LCA in the forestry area, there is a demonstration of the applicability of LCA in its breadth and diversity, with the works of Brugnara (2001), Mastella (2002); Silva (2012); Haaren et al. (2010); Morris et al. (2011); Zhang (2012); CANADA (2014); Reichert and Mendes (2014); VALINHOS (2011).

It can be seen that no Brazilian studies were found on greenhouse effect emissions according to the use and destination of urban tree waste. Therefore, as anthropogenic actions in recent decades and environmental impacts have significantly altered the climate and ecosystem dynamics on Earth. It is necessary to maintain environmental quality and make better use of available resources in the face of energy alternatives. Therefore, this study quantified the environmental impacts and analyzed four scenarios (landfill, landfill with methane recovery, reuse and incineration) for urban tree waste in Joao Pessoa.

3.2 MATERIALS AND METHODS

3.2.1Object of study

Information on the mass of urban forestry waste was provided by the Management Information Division of the Special Municipal Urban Cleaning Authority (EMLUR).

3.2.2Life Cycle Assessment

Life Cycle Assessment (LCA) is structured and standardized by the International Organization for Standardization (ISO), through ISO 14040 (2006) and ISO 14044 (2006). ISO 14040 (2006) shows that LCA is a technique that helps to analyze and interpret environmental impacts by surveying and compiling the inputs, production stages, consumption and outputs of a product system throughout its life cycle. In Brazil, LCA is governed by the ABNT NBR 14040 and 14044 standards (Brazilian Association of Technical Standards - ABNT, 2014a; ABNT, 2014b).

LCA consists of four main interrelated phases (KLOPFFER, 2012): Definition of the objective and scope of the analysis; Inventory of the processes involved, with the definition of the system's inputs and outputs; Analysis of the environmental impacts linked to the system's inputs and outputs; Interpretation of the results of the inventory and evaluation phases. A full explanation of LCA can be found in Guinèe (2001) and Guinèe (2002).

3.2.2.1 Defining the objective and scope of the analysis

Urban tree waste is collected by EMLUR on weekdays and, in exceptional and emergency cases, on weekends. This waste is collected by specific teams, without mixing it with other waste, and sent to the Joao Pessoa metropolitan landfill. At the landfill, they are weighed and sent for disposal in the waste cells.

The stage of waste transportation is the distance traveled by the truck between the city and the landfill. This information was obtained using a digital map on *Google Earth, taking into* account the route between the city center (departure) and the landfill (arrival). The route taken by the collection team was recorded by the drivers in terms of kilometers traveled (km).

3.2.2.2 Inventory of the processes involved, defining the inputs and outputs of the system

There is currently a lot of software available that has databases for processes and impact assessment, carrying out analysis and comparisons of complex product life cycles (GALDIANO, 2006). SimaPro 8.2.0.0 (SIMAPRO, 2016) is highly specialized software for developing LCA, which has thousands of processes and a series of methods for assessing the environmental impacts of products and services. The basic structures for analyzing environmental impacts are: characterization, damage assessment, normalization, weighting and addition of environmental damage (SIMAPRO, 2016). The database used was Ecoinvent, as it has data for thousands of products and processes (ECOINVENT, 2015) and is the most widely used in academic studies.

Due to current concerns about climate change, the Global *Warming Pot*ential (GWP) was selected as a measure of environmental impact, expressed in kilograms of carbon dioxide equivalent (kg $CO_{2\text{-}eq}$), thus characterizing the carbon footprint. The main body for assessing climate change is the Intergovernmental Panel *on Climate* Change (IPCC). The IPCC was created in 1988 by the United Nations Programme and the World Meteorological Organization to provide a scientific overview of climate change and possible environmental repercussions (IPCC, 2015).

In SimaPro, the method used was the IPCC 2013 GWP 100a (SIMAPRO, 2016). The GWP of any substance is the ratio between the contribution to the absorption of radiation heat resulting from the instantaneous discharge of 1 kg of a greenhouse gas and equal carbon dioxide (CO_2) emissions integrated over time (in this case, 100 years) (FERREIRA, 2004; IPCC, 2013).

3.2.2.2.1 Processes used for LCA

a) Urban tree waste

The urban tree residue modeled was a combination of five processes to express the variety of trees. The following processes were used (ECOINVENT, 2015):

- *Bark chips, wet, measured as dry mass {GLO}* ;
- *Cleft timber, measured as dry mass {GLO};*
- *Residual hardwood, wet {GLO}*;
- *Residual softwood, wet {GLO}*;
- *Wood chips and particles, willow {GLO}.*

b) Transportation

The transportation stage was divided into two stages. The first corresponds to the route taken by the vehicle when collecting waste within the urban perimeter, an average of 35 km/day between collection sites. The second corresponds to the route taken by the trucks from the city to the landfill, an average of 25 km to the disposal site.

For both moments, the *Transport, freight, lorry 7.5-16 metric ton, EURO3 {GLO}| market for | AllocDef, S* (ECOINVENT, 2015) process was used.

3.2.3Analysis of the environmental impacts associated with the system's inputs and outputs

The end-of-life treatments (final disposal scenarios) selected comprised four options for urban tree waste from the municipality of Joao Pessoa: landfill without and with methane collection, simple municipal incineration and wood reuse.

To model the landfill, it was assumed that the wood deposited decomposes slowly and forms methane and carbon dioxide during the first 150 years. Approximately 20% does not decompose and will remain in the landfill as stable material. Disposing of the waste in a landfill, without collecting the methane, is the disposal currently applied by the municipal authorities.

Some landfills are equipped with a methane collection system as a second final disposal option, where the methane formed is collected and used as fuel. It could also be considered that the methane is simply burned, since it is much better to emit CO_2 than methane. However, here it was considered that the methane was used as fuel, avoiding the emissions necessary for the production of natural gas. For every 1 kg of wood deposited in a landfill with methane collection, 0.007 kg of natural gas does not need to be produced. Air emissions include unused methane (0.002 kg) and total CO_2 emissions (0.5 kg), following the procedure described by PRé Consultants (2014).

Municipal incineration includes emissions from incineration and consumables for treating the combustion gas. The incinerator itself is also included in the process and the wood ash is considered to be landfilled.

The reuse of the wood means that the tree waste is shredded and sold, preventing new wood from being produced in this format. This scenario complies with the city of Joâo Pessoa's Integrated Solid Waste Management Plan for the reuse of urban tree material.

3.2.4Interpreting the results of the inventory and evaluation phases

The results were interpreted by quantifying the carbon footprint (kg CO_{2-eq}), in tons (t) for 2008, for each of the proposed scenarios.

The carbon footprints were calculated over 13 years (2003 - 2015) for all scenarios. The carbon footprint per ton of waste collected in each proposed scenario was also analyzed to facilitate comparison with existing scientific literature.

3.3 RESULTS AND DISCUSSION

Brazil is a signatory to several agreements on climate change, but has no binding commitment to reduce CO_2 emissions. The municipality of Joao Pessoa has plans to become a Sustainable City, through the Joao Pessoa Sustainable Action Plan (JOÂO PESSOA, 2014). The interesting thing is to show which alternatives are available and which have the greatest impact, in order to help public authorities make decisions and guide policies.

Figure 3.1 shows the evolution of urban tree maintenance waste generation (10^3 t) in Joao Pessoa from 2003 to 2015. The amount of total waste generated annually varied over the 13 years, depending on demand and urban planning.

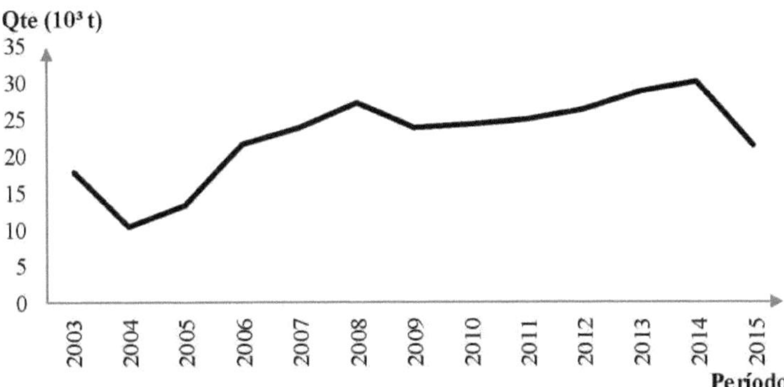

Figure 3.1. Evolution of urban tree maintenance waste generation (103 t) in Joao Pessoa, 2003 - 2015.

Historical data on the collection of forestry waste indicates an increase in collection over time. This could be an indication of efficient waste management. The smallest amount of waste was collected in 2004, 10,139 tons, and the largest amount was collected in 2014, 30,023.87 tons.

Figure 3.2 shows the results of the LCA processes for the four scenarios (landfill, methane utilization, incineration and reuse) for urban forestry waste in Joâo Pessoa in 2008 (27,065.04 t of waste collected).

Figure 3.2.a) shows the scenario of disposing of urban forestry waste in the landfill. Emissions from tree waste simply deposited in the landfill generated 3,690 t CO_2 -eq, causing the greatest impact on the environment of all the options examined. The results also indicate that the landfill process, excluding the transportation process, emits 1,870 t co2-eq.

Figure 3.2.b) corresponds to the landfill scenario with methane collection, where it was found that this process releases 3,070 t co2-eq, demonstrating a reduction compared to the simple landfill scenario. The methane collection process at the landfill prevented the generation of 620 t CO_2 -eq, reducing GHG emissions compared to the previous process.

Landfilla) Use of methaneb)

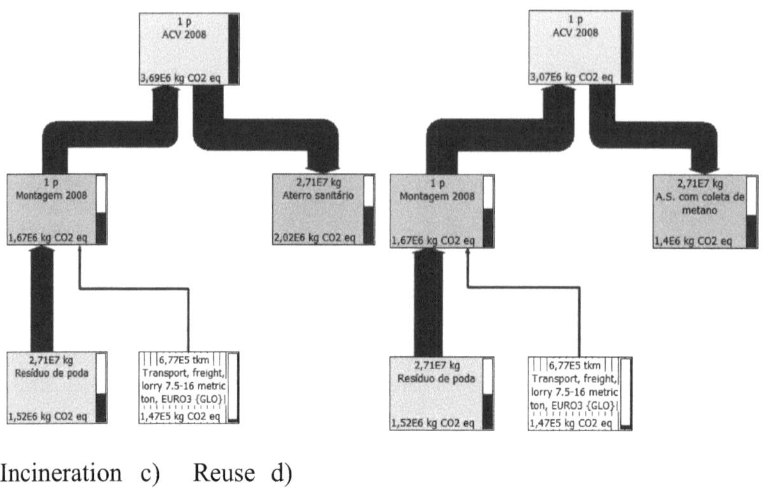

Incineration c) Reuse d)

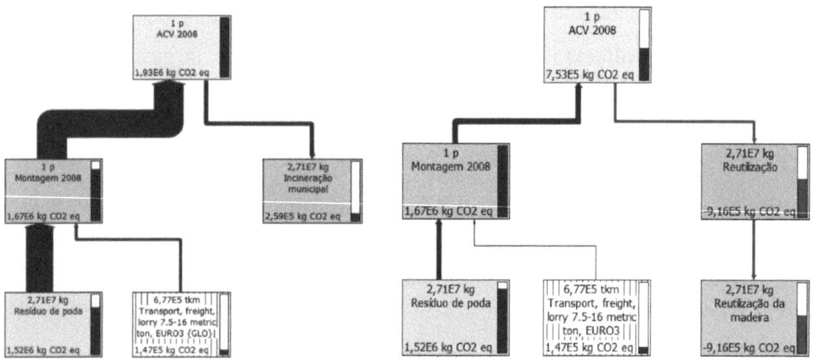

Figure 3.2. Quantification of four LCA scenarios for urban tree maintenance waste in the city of Joao Pessoa.

Figure 3.2.c) represents the municipal incineration scenario, where the result indicated emissions of 1,930 t CO_2 -eq. The process of incinerating tree waste (without using heat or electricity) generated 259 t of GHG.

Figure 3.2.d) shows the reuse scenario, the process with the lowest environmental impact, emitting 753 t $_{CO2\text{-eq}}$. The reuse process was negative, emitting 916 t $_{CO2\text{-eq}}$.

It can be seen that the processes of collecting and transporting forestry waste are common to all scenarios, emitting the same amount of GHG, 1,670 t $_{CO2\text{-eq}}$.

With regard to the more specific processes of each scenario, there is a change in GHG emissions, starting with a negative balance and ending with a positive balance, a behavior caused by the externality. In the most impactful scenario, emissions amounted to 3,690 t $_{CO2\text{-}eq}$, gradually decreasing and ending with no emissions of 753 t $_{CO2\text{-}eq}$ (positive).

The scenario with the highest GHG emissions was simple landfill disposal, which is currently used by the municipal authorities. The waste reuse scenario proved to be the best option, as it avoids producing "virgin" briquettes by reusing tree waste.

The result of the CO_2 -eq emissions per ton of urban forestry waste showed that the landfill scenario emits 136.34 kg of CO_2 /t collected. For the methane collection and municipal incineration scenarios, the study found emissions of 113.43 and 71.31 kg of CO_2/t, respectively. The reuse scenario had the lowest environmental impact, 27.82 kg of $_{CO2/t}$ collected.

Figure 3.3 shows the behavior of the carbon footprint ($_{103}$ t CO_2 -eq) of the disposal scenarios for urban tree waste in Joao Pessoa, in the period 2003 and 2015. The current scenario (simple sanitary landfill) employed by the municipal government had an accumulated release of 39,802 t CO_2 -eq over a 13-year period. These gases are produced during the waste decomposition process by decomposing microorganisms.

It can be seen that the public authorities in Joao Pessoa use the most polluting (with the highest carbon footprint) and least noble form of waste disposal, and that the municipal government needs to pay more attention to making better use of this material. It can be seen that as urban forestry waste is better disposed of, GHG emissions decrease.

For the methane collection scenarios, the GHG emitted over the period was 33,134 t, lower than the current system, and higher than that emitted by the municipal incineration process, 20,830 t $_{CO2\text{-eq}}$.

If all the biomass generated by the maintenance of Joao Pessoa's urban tree plantations were reused in the form of briquettes between 2003 and 2015, the emissions avoided would be 8,126 t CO -eq.$_2$

The reuse of wood has become more advantageous in terms of environmental impact than disposal in the metropolitan landfill (current scenario), with an accumulated difference of 31,676 t CO_2 -eq over 13 years.

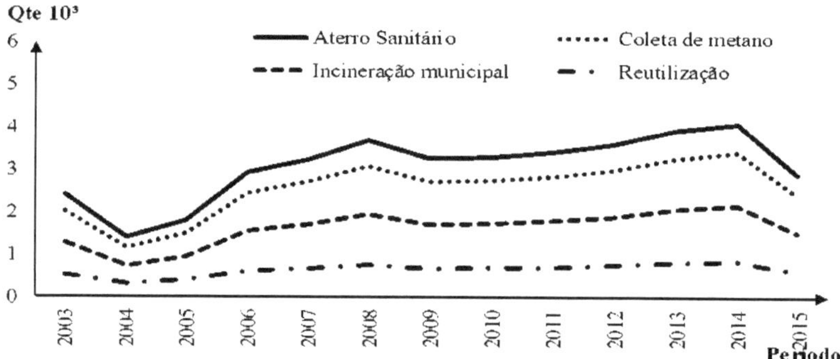

Figure 3.3. Evolution of the carbon footprint (10^3 t CO_2 -eq) for the four urban tree maintenance waste disposal scenarios for Joao Pessoa/PB, from 2003 to 2015.

Even though the city is in a relatively comfortable situation as far as air quality is concerned, trends in economic growth and consumption are putting the effects of GHGs in the city on alert. The GHG emissions inventory carried out in 2010 concluded that 1,198,034 t $_{CO2\text{-eq}}$ were emitted in that year, with an increase of 43.70% for 2012, generating 1,721,681 t CO2-eq (JOÂO PESSOA, 2014).

The disposal of urban forestry waste used by the municipality (landfill) emitted 3,274 t CO_2 -eq and 3,578t CO_2 -eq in 2010 and 2012, respectively. In 2010, the GHG emitted in the disposal of urban trees corresponded to 0.27% of the total released in the city, and in 2012, they corresponded to 0.21% of the total emission. For the reuse scenario (the best scenario studied), emissions corresponded to 0.06% of the gases emitted in the city in 2010, and 0.04% of the gases released in 2012 in the municipality.

With these comparisons, it can be seen that the GHGs generated by urban forestry waste in the landfill correspond to a small percentage in relation to the total emitted in the respective years, but could be even lower if the biomass were destined for reuse. Even so, because these gases accumulate over time, their emission contributes to altering the city's microclimate, where the change in disposal used by the municipality represents a reduction in GHG emissions.

The Sustainable Joào Pessoa Action Plan presents among the strategies and actions to control GHG, the use of biomass in the energy matrix and the promotion of waste recycling. A strong ally in meeting the targets proposed by the municipality is the use of waste from the maintenance of urban trees (JOÂO PESSOA, 2014).

In 2001, logging in the Amazon rainforest, using the traditional model, released around 147,790 tons of co2/year into the atmosphere. In an exploratory model, using appropriate techniques and mitigation of environmental impacts, 34,228 t co2/year were released for the entire production chain (BRUGNARA, 2001).

When comparing the results found by Brugnara (2001) and those obtained in this study, in 2014, the emissions caused by the disposal used by the municipality (landfill) correspond to the same impact emitted by the exploitation of 163.91 ha of native vegetation for timber purposes, in the traditional model. The scenario found in the study in 2014, reuse of forestry waste, corresponds to the exploitation of 65.79 ha of Amazon forest in the traditional model.

Therefore, the impacts generated by the reuse of waste correspond to the exploitation of smaller areas of native vegetation in the Amazon rainforest, when compared to the current destination (landfill), used by the public authorities.

The conclusion of Morris et al. (2011) in Canada is the opposite of that found in the municipality of Joâo Pessoa, where gas recovery and incineration were the scenarios that caused the highest GHG emissions. Corroborating the results obtained here, (Mendes *et al.* 2004 *apud* CABRAL, 2010) state that the use of landfills has a greater environmental impact than incineration, *since* it contributes considerably to global warming (methane emissions). Even when landfills have a system for collecting and burning biogas, approximately 60% is lost to the environment (DASKALOPOULOS *et al.* 1998 *apud* CABRAL, 2010) and what can be collected has a lower calorific value than natural gas (HAMER, 2003 *apud* CABRAL, 2010).

One of the solutions for minimizing risks to human health and the environment is through proper waste management (LEVIS; BARLAZ, 2013). The management of urban tree maintenance waste has been the subject of attention because it contains recyclable materials, has the potential for energy use and contributes to environmental sustainability.

3.4 CONCLUSIONS

From the analysis carried out, it was concluded that:

The current disposal system used by the municipal authority (landfill) is the most environmentally damaging of all the scenarios studied. With emissions of 136.34 kg of CO_2 -eq/t of waste, it has the highest carbon footprint in the study;

The best disposal options, with the lowest GHG emissions, were the reuse of tree waste, which had the lowest environmental impact, and the incineration scenario;

The destination proposed by the Municipal Solid Waste Plan (reuse) is an environmentally ideal and less harmful alternative, to be implemented in the short term;

The study showed that the reuse of biomass, as well as being environmentally viable, has the potential to contribute to the city's environmental quality, including the possibility of being used as carbon credits.

3.5 REFERENCES

BRAZILIAN ASSOCIATION OF TECHNICAL STANDARDS - CATALOG. **ABNT NBR ISO 14044: 2009 Corrected Version: 2014.**2014a. Available at: <https://www.abntcatalogo.com.br/norma.aspx?iD=316461>. Accessed on: January 3, 2017.

ABNT NBR ISO 14040: 2009 Corrected Version: 2014. 2014b. Available at: <https://www.abntcatalogo.com.br/norma.aspx?iD=316461>. Accessed on: January 3, 2017.

ABRELPE. BRAZILIAN ASSOCIATION OF PUBLIC CLEANING COMPANIES. **Overview of**

solid waste in Brazil: 2014. Sao Paulo. 2015. Available at: <http://www.abrelpe.org.br/Panorama/panorama2014.pdf>. Accessed on: November 13, 2017.

AVALIAÇÂO DO CICLO DE VIDA BRASIL - ACV BRASIL. **Life Cycle Assessment**. Available at: <http://acvbrasil.com.br/avaliacao-do-ciclo-de-vida/>. Accessed on: February 18, 2017.

BARATTA JUNIOR, A. P.; MAGALHÂES, L. M. S. Utilization of tree pruning waste from the city of Rio de Janeiro for composting. **Revista de Ciências Agro- Ambientais**, Alta Floresta, v. 8, n.1, 2010.

BRUGNARA, G. A. **Forests, wood and housing:** energy and environmental analysis of the production and use of wood as a contribution to the challenge of valuing the Amazon Forest. 2001. Dissertation (Master's Degree in Energy Systems Planning).

State University of Campinas, Faculty of Mechanical Engineering, Campinas, 2001.

CABRAL, E. **Considerations on Solid Waste.** Federal University of Ceara, Fortaleza. Class material. Available at:

<http://www.deecc.ufc.br/Download/Gestao_de_Residuos_Solidos_PGTGA/CONSIDER ACOES_SOBRE_RESIDUOS_SOLIDOS.pdf>. Accessed on: February 21, 2017.

CANADA. Information Center Government of Alberta. **Recommendations for Reducing Leaf and Yard in Aberta**. Alberta Environment and Sustainable Resource Development, 2014. Available at: <http://aep.alberta.ca/waste/documents/ReducingLeafYardWaste-Feb2014.pdf>. Accessed on: Feb. 10, 2017.

CARVALHO, M.; FREIRE, R. S.; MAGNO, A. H. **Promotion of sustainability by quantifying and reducing the carbon footprint:** new practices for organizations. In: Global Conference on Global Warming, 2015, Athens, Greece. Proceedings of the global conference on global warming, 2015.

ENERGY COMPANY OF MINAS GERAIS. **Afforestation manual**. Belo

Horizonte: MG, p. 112, 2011. Available at:

<http://www.cemig.com.br/sites/imprensa/pt-br/Documents/Manual_Arborizacao_Cemig_Biodiversitas.pdf>. Accessed on: January 10, 2017.

PARANAENSE ENERGY COMPANY. **Arborization of public roads**. Curitiba: PR, 2009. Available at:

<http://www.copel.com/hpcopel/guia_arb/copel_e_a_arborizacao_de_vias_publicas.html>.

Accessed on: February 15, 2017.

ECOINVENT CENTER. **What we do**. Zurich, 2015. Available at: <http://www.ecoinvent.org/about/about.html>. Accessed on: December 16, 2016.

EMLUR. Special Municipal Cleaning Authority of Joao Pessoa. **Information on annual pruning and costs**. Joao Pessoa. 2016.

EMLUR. Special Municipal Cleaning Authority of Joao Pessoa. **Municipal Plan for Integrated Solid Waste Management (PMGIRS)**: Diagnosis and Planning of Urban Cleaning and Solid Waste Management Services. Joao Pessoa. v. 2, 2014. Available at:< http://transparencia.joaopessoa.pb.gov.br/dadospublicos/?p=111>. Accessed on: January 13, 2017.

FERREIRA, J. V. R. **Product Life Cycle Analysis**. Viseu: Polytechnic Institute of Viseu, 2004.

FREIRE, R. S.; CARVALHO, M.; CARMONA, C. U. M.; MAGNO, A. H. **Perspectives on the implementation of climate change public policies in Brazil.** In: Global Conference on Global Warming, 2015, Athens, Greece. Proceedings of the global conference on global warming, 2015.

GALDIANO, G. P. **Life Cycle Inventory of Offset Paper Produced in Brazil**. 2006. Dissertation (Master's in Engineering). University of Sao Paulo, Sao Paulo, 2006.

GOLVEIA, N. Residuos sólidos urbanos: impactos socioambientais e perspectiva de manejo sustentàvel com inclusao social. **Ciência & Saùde Coletiva**, v. 17, n. 6, p. 1503:1510, 2012.

GUINÉE, J. B. (ed) **Life Cycle Assessment: An operation al guide to the ISO Standards;** LCA in Perspective; Guide; Operational Annexto Guide. Center for Environmental Science, Leiden University, The Netherlands, 2001.

GUINÉE, J. B. **Hand book on life cycle assessment:** operational guide to the ISO standards**.** Kluwer Academic Publishers, Boston, 2002.

HAAREN, R. van; THEMELIS, N. J.; BARLAZ, M. LCA comparison of windrow composting of yard wastes with use as alternative daily cover (ADC). **Waste Management**, v. 30, f 2649-2656, 2010.

INTERGOVERNMENTAL PANEL ON CLIMATE CHANGE. **What is the IPCC?** Geneva: Secretariat of the WMO, 2013. Available at: <http://www.ipcc.ch/>. Accessed on: January 1, 2017.

. **History**. Geneva: Secretariat of the WMO, 2015. Available at: <http://www.ipcc.ch/organization/organization_history.shtml>. Accessed on: February 1, 2017.

JOÂO PESSOA. Joao Pessoa City Hall. Joao **Pessoa Sustainable Action Plan 2014.** Joao Pessoa, 2014. Available at:

<http://www.joaopessoa.pb.gov.br/>. Accessed on: April 3, 2017.

KLOPFFER, W. **The critical review life cycle assessment studies according to ISO 14040 and 14044:**origin, purpose and practical performance. The International Journal Of Life Cycle Assessment, Heidelberg, p. 1-7. 2012.

LEVIS, J. W.; BARLAZ, M. A. **Composting Process Model Documentation**; Project Report, North Carolina State University: Raleigh, NC. 2013 Available at: <http://www4.ncsu.edu/~jwlevis/Composting.pdf>Accessed Feb. 11, 2017.

MARTINS, C. H. The use of wood from road tree pruning in Maringa/PR. **Revista Verde**, Mossoró, v. 8, n. 2, p. 257 - 267, 2013.

MASTELLA, D. V. **Comparison between the production processes of ceramic and concrete blocks for structural masonry, through life cycle analysis**. 125 f., 2002. Dissertation (Master's degree in Civil Engineering). Federal University of Santa Catarina, Florianópolis, 2002.

MEIRA, A. M. **Waste management in urban forestry**. 2010. 179 f Thesis (Doctorate in Sciences). Luiz de Queiroz College of Agriculture, University of São Paulo, Piracicaba, 2010.

MORRIS, J.; MATTHEWS, H. S.; MORAWSKI, C. **Review of LCAs on Organics Management Methods & Development of an Environmental Hierarchy**. Information Center Alberta Enviroment. Alberta, 2011. Available at:<http://environment.gov.ab.ca/info/library/8350.pdf>. Accessed on: Feb. 10, 2017.

PIRES, A.C; RABELO, R. R; XAVIER, J. H. V. **Potential use of Life Cycle Analysis (LCA) associated with organic production concepts applied to family farming**. Cadernos de Ciência e Tecnologia, Brasilia, v. 19, n. 2, p. 149-178, 2002.

PRé. SimaPro Tutorial. [s.l.]: [s.n.], 2014.

SÂO PAULO (Capital). Municipal Department of Greenery and the Environment. **Technical Manual for Urban Afforestation**. Sao Paulo, 2015. Available at:

<http://www.prefeitura.sp.gov.br/cidade/secretarias/meio_ambiente/publicacoes_svma/ind ex.php?p=188452>. Accessed on: January 07, 2017.

NATIONAL SECRETARIAT FOR ENVIRONMENTAL SANITATION. National Sanitation Information System. **Diagnosis of urban solid waste management - 2014.** Part 2-Table of Information and Indicators. Brasilia: MCIDADES. SNSA, 2016

SILVA, D. A. L. **Life cycle assessment of MDP wood panel production in Brazil**. 2012. Dissertation (Master's Degree in Materials Science and Engineering).

University of Sao Paulo, Sao Carlos School of Engineering, Sao Carlos, 2012.

SIMAPRO. **SimaPro Data base 8**: Manual Methods Library. California, 2016. Available at: <https://www.pre-sustainability.com/simapro-database-and-methods-library>. Accessed on: April 4, 2017.

RECIFE. Department of the Environment and Sustainability. **Manual de arborização:** orientaçao e manutença da arborização urbana da cidade do Recife. Ed. Recife: [S.N.]. 2013. Available at:<ww2.recife.pe.gov.br/wp- content/uploads/Manual_Arborizacao.pdf>. Accessed on: April 13, 2017.

REICHERT, G. A.; MENDES, C. A. B. Life cycle assessment and decision support in integrated management and sustainability of urban solid waste. **Eng. Sanit. Ambiental**, v. 19, n. 13, 2014.

VALINHOS. Valinhos City Hall. **Integrated Solid Waste Management Plan for the Municipality of Valinhos-SP**. Valinhos, 2011. Available at: <http://www.valinhos.sp.gov.br/portal/arquivos/planejamento/PGIRS_- _Verso_Preliminar_2.pdf>. Accessed on: February 10, 2017.

ZHANG, S.**What is the Hisghest and Best Use** of **Organic Solid Waste:** Production of Compost and Production of Energy?.Greenest City Scholar, 2012.Available at:

<https://sustain.ubc.ca/sites/sustain.ubc.ca/files/Zero%20Waste%20%20-%20Siduo%20Zhang%20-%20Compost%20vs%20%20Energy%20Production.pdf>. Accessed on: February 10, 2017.

4. ARTICLE 3 - ELECTRICITY GENERATION AS A CLEAN DEVELOPMENT STRATEGY FOR WASTE FROM THE URBAN ARBORIZATION OF JOO PESSOA

YURI ROMMEL VIEIRA ARAÙJO; MONIJANY LINS GÓIS; LUIZ MOREIRA COELHO JUNIOR; MONICA CARVALHO

SUMMARY

This article assessed the environmental impacts of three scenarios (landfill, electricity generation and heat generation) for urban tree waste in Joao Pessoa, from a clean development perspective. The IPCC 2013 GWP 100a was used in the Life Cycle Assessment (LCA) to quantify the environmental impacts of the disposal scenarios for urban tree maintenance. Based on the analysis carried out, the main conclusions were that: The current disposal method used by the municipal authority (landfill) has the greatest impact on the environment, with emissions of 136.34 kg of $CO_{2\text{-eq/t}}$; the generation of electricity had a positive externality and proved to be more environmentally viable than the generation of heat; the implementation of a clean development mechanism project at the landfill, for the use of urban tree waste, is a strategy for the management of solid urban waste.

Keywords: environmental impact, greenhouse gases, renewable energies.

4.1 INTRODUCTION

Municipal solid waste (MSW) originates from domestic activities and urban cleaning (ABRELPE, 2015). Due to the complexity and scope of MSW management, there are difficulties in most Brazilian municipalities. Despite all efforts, the proper management of urban tree waste is not always a priority on municipal agendas (MEIRA, 2010).

Urban afforestation is divided into green areas (parks, woods, squares and gardens) and street afforestation (public roads). Maintenance in these areas is necessary and includes irrigation, complementary fertilization, preventive treatments, pruning and replacement of individuals (COPEL, 2009).

Urban pruning involves removing branches, fruit, inflorescences or foliage in order to promote the longevity of the tree. Tree waste represents a significant portion of urban waste, which is deposited in landfills and dumps, and in some cases incinerated (BARATTA JUNIOR; MAGALHÂES, 2010; CEMIG, 2011; RECIFE, 2013).

The disposal of forestry waste results in a high cost for Brazilian municipalities. This means a waste of resources, since tree-planting waste has energy potential in various industrial sectors (MARTINS, 2013). In 2014, in the municipality of Belo Horizonte (MG), urban tree waste accounted for 9.37% of the solid waste sent to processing plants. In Igaracy (PB), 7.10% of municipal solid waste is made up of biomass from urban trees (SNSA, 2016). In 2010, in the state of São Paulo, disposing of urban tree waste in a private landfill cost an average of R$ 68.00/t (MEIRA, 2010). In Joâo Pessoa (PB) the costs are on average R$ 200.00/t, including transport, fuel, labor, landfill disposal and other operating costs (EMLUR, 2016).

Camilo et al. (2008) studied 70 municipalities in the state of São Paulo and found that there are no municipal regulations on the management of waste from the maintenance and removal of urban trees. It was clear that cities do not have guidelines and responsibilities for tree care and pruning (CAMILO et al, 2008).

In the municipality of Joao Pessoa (PB), between 2008 and 2014, the total accumulated waste from tree planting was 127x103 tons. According to the Municipal Plan for the Integrated Management of Solid Waste, by 2022 the waste from tree planting should be sent to a sorting center for reuse. The material with energy potential will be transformed into briquettes and the non-energy material will be composted. The Special Municipal Urban Cleaning Autarchy (EMLUR) is responsible for promoting the implementation of processing units by the private sector (EMLUR, 2014).

Improper disposal of MSW has an immediate impact on the environment and health, as well as contributing (through atmospheric emissions) to climate change. The greatest impact occurs during the decomposition stage, with the potential for contamination of the soil and surface and groundwater, the formation of toxic, asphyxiating and explosive gases, and the generation of greenhouse gases (GHG), mainly methane (CH_4) (GOLVEIA, 2012). These environmental impacts can be quantified by developing a Life Cycle Assessment (LCA) of the waste.

Life Cycle Assessment (LCA) has become increasingly important in measuring the environmental impacts of activities and processes, as it stems from the awareness that improvements in a given process can induce secondary effects throughout the life cycle, which positively and/or negatively affect the environmental performance of goods and services (CARVALHO et al., 2015; FREIRE et al., 2015; PIRES et al., 2002). The life cycle is understood as the stages necessary for a product or service to be developed or designed, fulfill its respective function and reach the stage of disposal, recycling or reuse.

LCA is therefore a methodology capable of assessing environmental impacts from the extraction of natural resources to the final disposal of the product (GUINÉE, 2001; GUINÉE, 2002). LCA can be used, for example, for: product development and improvement; defining strategic planning and public policies; managing the environmental impacts of products and services and responsible ecological marketing (ACV BRASIL, 2015; CARVALHO et al., 2015; FREIRE et al., 2015).

The applicability of LCA is wide and diverse, and the Integrated Solid Waste Management Plan for the Municipality of Valinhos (SP) mentions that permanent research into the characterization and LCA of products should be increasingly encouraged to support Integrated Management, also focusing attention on solid waste management policies, with shared responsibility between society, the productive sector and public authorities (VALINHO, 2011).

After an exhaustive and detailed systematic review, no Brazilian studies were found on the management of urban tree waste and its possibilities for inclusion in Clean Development Mechanisms (CDM). This article assessed the environmental impacts and analyzed three scenarios (electricity generation, heat generation and landfill disposal) for urban tree waste in Joao Pessoa, with a view to its inclusion in the Clean Development Mechanism (CDM).

4.2 MATERIALS AND METHODS

4.2.1Object of study

The information on the mass of urban forestry waste was provided by the Management Information Division of the Special Municipal Urban Cleaning Authority (EMLUR).

4.2.2Life Cycle Assessment

Life Cycle Assessment (LCA) is structured and standardized by the International Organization for Standardization (ISO), through ISO 14040 (2006) and ISO 14044 (2006). ISO 14040 (2006) shows that LCA is a technique that helps to analyze and interpret environmental impacts by surveying and compiling the inputs, production stages, consumption and outputs of a product system throughout its life cycle. In Brazil, LCA is governed by the ABNT NBR 14040 and 14044 standards (Brazilian Association of Technical Standards - ABNT, 2014a; ABNT, 2014b).

LCA consists of four main phases that are interrelated (KLOPFFER, 2012): Definition of the objective and scope of the analysis; Inventory of the processes involved, with the definition of the system's inputs and outputs; Analysis of the environmental impacts linked to the system's inputs and outputs; Interpretation of the results of the inventory and evaluation phases. A full explanation of LCA can be found in Guinèe (2001) and Guinèe (2002).

4.2.2.1 Defining the objective and scope of the analysis

Waste from urban tree planting in Joâo Pessoa/PB is collected by EMLUR on weekdays and, in exceptional and emergency cases, at weekends. This waste is collected by specific teams, without mixing it with other waste, and sent to the Joâo Pessoa Metropolitan Landfill. At the landfill, they are weighed and sent for disposal in the waste cells.

The stage of waste transportation is the distance covered by the truck between the city and the landfill. This information was obtained using a digital map on *Google Earth, taking into* account the route between the city center (departure) and the landfill (arrival). The route taken by the urban tree maintenance waste collection team was recorded by the drivers in terms of kilometers traveled (km).

4.2.2.2 Inventory of the processes involved, defining the inputs and outputs of the system

There are several software programs available today that have databases for processes and impact assessment, carrying out analysis and comparisons of complex product life cycles (GALDIANO, 2006). SimaPro (SIMAPRO, 2015) is highly specialized software for developing LCA, which has thousands of processes and a series of methods for assessing the environmental impacts of products and services. The basic structures for analyzing environmental impacts are: characterization, damage assessment, normalization, weighting and addition of environmental damage (SIMAPRO, 2015). The database used was *Ecoinvent*, within Simapro version 8.2.0.0, which has data for thousands of products and processes (ECOINVENT, 2015), as well as being the most widely used in academic studies.

Due to current concerns about climate change, the Global *Warming Pot*ential (GWP) was selected as a measure of environmental impact, expressed in kilograms of carbon dioxide equivalent (kg $CO_{2\text{-}eq}$), thus characterizing the carbon footprint. The main body for assessing climate change is the Intergovernmental Panel *on Climate* Change (IPCC). The IPCC was created in 1988 by the United Nations Programme and the World Meteorological Organization to provide a scientific overview of climate change and possible environmental repercussions (IPCC, 2015).

In SimaPro, the method used was the IPCC 2013 GWP 100a (SIMAPRO, 2015). The GWP of any substance is the ratio between the contribution to the absorption of heat from radiation resulting from the instantaneous discharge of 1 kg of a greenhouse gas and equal emissions of carbon dioxide (CO_2) integrated over time (in this case, 100 years) (FERREIRA, 2004; IPCC, 2013).

4.2.2.2.1 Processes used for LCA

The processes used for the LCA were urban forestry waste and transportation, which are described below:

a) Urban forestry waste

The modeled residual was a combination of five processes to express the variety of trees. The following processes were used

(ECOINVENT, 2015):

- *Bark chips, wet, measured as dry mass {GLO} ;*
- *Cleft timber, measured as dry mass {GLO};*
- *Residual hardwood, wet {GLO};*
- *Residual softwood, wet {GLO};*
- *Wood chips and particles, willow {GLO}.*

b) Transportation

The transportation stage was divided into two stages. The first corresponds to the route taken by the vehicle to collect the waste within the urban perimeter, an average of 35 km/day between collection points. The second corresponds to the route taken by the trucks from the city to the landfill, an average of 25 km to the disposal site.

For both moments, the *Transport, freight, lorry 7.5-16 metric ton, EURO3 {GLO}| market for | AllocDef, S* (ECOINVENT, 2015) process was used.

4.2.2.3 Analysis of the environmental impacts associated with the system's inputs and outputs

The end-of-life treatments (final disposal scenarios) selected comprised three options for urban tree waste from the municipality of Joao Pessoa: disposal in the landfill (with no use of methane, the current situation), use to generate electricity and use to generate heat.

Wood deposited in landfills decomposes slowly and forms methane and carbon dioxide during the first 150 years.

Approximately 20% does not decompose and remains in the landfill as stable material. The disposal of urban forestry waste in the landfill, without the collection of methane, is the disposal currently applied by the municipal authorities. The inventory for this scenario includes the landfill itself, but also the treatment of the leachate[1] in the first 100 years. The decomposition of waste after 100 years does not generate atmospheric emissions, because at this point the methane production phase is over and no *landfill gas*[2] is produced.

Incineration with the use of electricity takes into account the construction of the incinerator itself, depositing the ashes in a landfill and all the emissions and treatments involved in incinerating the forestry waste. For electricity, 1.74 MJ/kg of waste was considered (PCI = 13.99 MJ/kg of tree waste).

The incineration of arboricultural waste with the use of heat takes into account the construction of the incinerator, depositing the ashes in a landfill, and all the emissions and consequent treatments for incineration. We considered 3.49MJ/kg for heat (PCI = 13.99 MJ/kg urban tree waste). In this case, for every kg of tree waste incinerated, 3.49MJ of heat from plant biomass is avoided.

4.2.2.4 Interpreting the results of the inventory and evaluation phases

The results were interpreted by quantifying the carbon footprint (kg $_{CO2\text{-}eq}$) in tons (t) for 2008, for each of the proposed scenarios. Another result analyzed was the carbon footprint per ton of waste collected in each proposed scenario, to facilitate comparison with existing scientific literature.

The carbon footprints were recorded over 13 years (2003 - 2015) for all scenarios.

4.3 RESULTS AND DISCUSSION

In order to avoid the large amount of pollution and minimize the anthropogenic greenhouse effect on the Earth, the search to maintain environmental quality has arisen in the face of the need to find energy alternatives. Three possibilities for using and/or disposing of tree-planting waste have been studied here, with a view to

1 Liquid effluent generated as a result of rainwater percolating through solid waste disposed of in landfills, as well as the natural humidity of this waste. Also known as leachate.

2 from the English *landfill gas (LFG).* In Portuguese, the terms "landfill gas" and "biogas" are used to refer to the gas produced by the decomposition of solid waste in landfills.

verifying their potential for mitigating environmental change. Although Brazil does not have a mandatory commitment to reduce GHG emissions, it is interesting to see which alternatives would have the greatest impact in this regard.

Figure 4.1 shows the mass (t) of urban tree waste collected by EMLUR in Joao Pessoa and taken to the Metropolitan Landfill between 2003 and 2015. There has been an increase in the generation of urban tree waste over the last 13 years. This growth may be the result of increased efficiency on the part of the municipal government in managing this waste, or the search on the part of the population to give it a more efficient and environmentally correct destination.

After specifying the amount of urban forestry waste in each scenario proposed above, Table 4.1 shows the results of implementing the LCA in the SimaPro software (2015) with the IPCC 2013 GWP 100a method (IPCC, 2013), for waste from the year 2008 in Joao Pessoa.

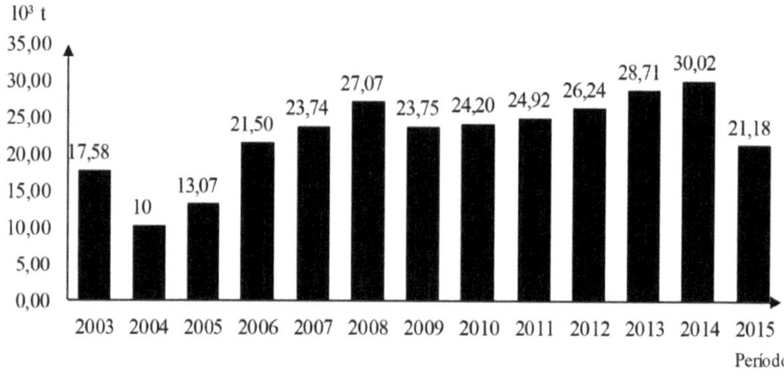

Figure 4.1. Evolution of the mass of urban forestry waste received at the Joao Pessoa Metropolitan Landfill from 2003 to 2015.

Table 4.1. *Carbon footprint associated with three disposal scenarios for tree-planting waste in Joao Pessoa/PB for the year 2008.*

	Carbon footprint		
Urban tree waste collected in 2008	Landfill	Incineration using electricity	Incineration with heat recovery

27.065,04 t	3,690 t CO2-eq	-4,360 t CO2-eq	1,432 t CO2-eq

- **Landfill:** Arboricultural waste simply deposited in the landfill generated a high carbon footprint, with the greatest impact on the environment of all the options considered, generating 3,690 tCO_2 -eq. Forestry waste and its transportation accounted for 1,670 t $_{CO2\text{-}eq}$ (approximately 52% of the final total), and uncontrolled decomposition in landfill contributed 2,020 t $_{CO2\text{-}eq}$.

- **Use for electricity generation:** This scenario went one step further than simple incineration, avoiding the consumption of electricity from the grid. It had the best results of all the scenarios studied, with emissions of -4,360t CO_2 -eq. Forestry waste and its transportation accounted for 1,670 t $_{CO2\text{-}eq}$, and electricity generation prevented the emission of 6,030 t CO_2 -eq. These final negative emissions indicate the possibility of climate change mitigation, as well as incorporation into Clean Development Mechanisms (CDM).

- **Use to generate heat:** This heat can be sold to industries or condominiums, or even used in the landfill's internal processes, avoiding the consumption of fuel in boilers for this purpose. This process emitted 1,432 t CO_2 -eq. Forestry waste and its transportation accounted for 1,670 t $_{CO2\text{-}eq}$, and the use of heat from incineration prevented 9450 MJ of heat from being generated through the incineration of plant biomass, avoiding the emission of 238 t $_{CO2\text{-}eq}$.

It was observed that the collection and transportation of tree waste are common processes for all disposal scenarios. The scenario with the highest tCO_2 -eq emissions was landfill disposal, an action currently employed by the municipal authorities. Using the material to generate electricity proved to be the best option for the waste, where the results indicate a negative balance, i.e. a significant amount of GHGs will no longer be emitted in the region, minimizing atmospheric environmental impacts. Table 4.2 shows the carbon footprint per ton of forestry waste processed.

Table 4.2. Carbon footprint per tonne (kg of CO_2 .eq/t) of waste from urban tree planting in the city of Joao Pessoa/PB, for each scenario studied.

Process	Emission (kg CO_2 eq/t waste collected)
Landfill	136,338
Electricity Generation	-161,093
Heat Generation	52,909

Considering that the municipal authorities currently dispose of urban tree waste in the Joao Pessoa metropolitan landfill, the best scenario studied for 2008 was electricity generation.

Carbon footprints were calculated for these two scenarios for the period 2003 - 2015.

Figure 4.2 shows the carbon footprints for the disposal scenarios between 2003 and 2015. The current scenario employed by the municipal authorities (simple landfill) had a cumulative release of 39,826 t CO_2 -eq over a 13-year period. These gases are produced during the waste decomposition process by micro-decomposing organisms.

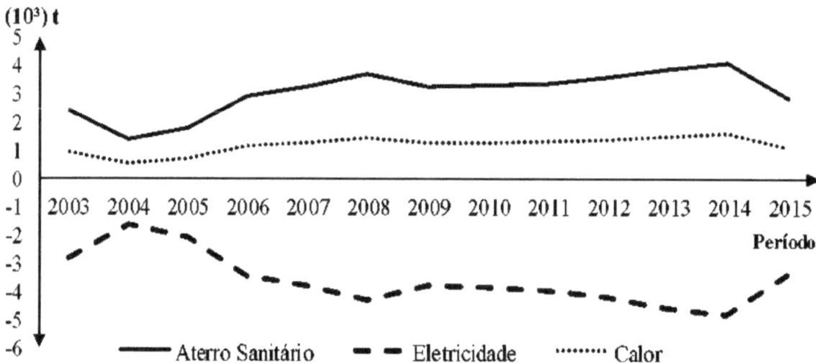

Figure 4.2. Carbon footprint for the scenarios of disposing of urban forestry waste in the landfill, and using the waste to generate electricity and heat, in the period 2003-2015, for Joâo Pessoa/PB.

When comparing the current scenario with the best-case scenario, it is clear that the public authorities are currently using the most environmentally damaging method. If all the biomass generated by urban forestry in Joâo Pessoa between 2003 and 2015 were used to generate electricity, it would prevent the emission of 47,058 t $_{CO2\text{-}eq}$ in the city.

Cortez (2011) mentioned that there are few reports of the use of tree waste on a large scale (whether for energy purposes or not), and lists reuse as firewood and charcoal production as disposal methods. The work by Morris, Matthews and Morawski (2011) reviewed various final disposal scenarios for *leaf and tree waste* from the municipality of RedDeer (Canada), concluding that landfill with gas recovery for energy was preferable to incineration with electricity production (considering climate change, human carcinogens and toxicity to the ecosystem).

Still with regard to the environmental impacts of using landfill gas, the US Environmental Protection Agency (EPA, 2015) points out that there are variations in the methane collection systems present in landfills. And that different technologies are commercially available for producing electricity from waste, which can significantly alter the results of the analysis carried out.

The study by *The Leaf and Yard Waste Diversion Technical Committee* (CANADA, 2014) indicates that there are better options for urban tree waste than landfill (composting, for example), mainly for economic reasons. Including not using space (land use), and for environmental reasons (reduction of GHG emissions).

With regard to using it to generate electricity, Cortez (2011) studied the potential for using urban tree waste to generate electricity. Developing technical, economic and environmental analyses for the steam cycle and anaerobic digestion, he concluded that a small thermoelectric plant was the most viable way of using this material.

In a similar vein, Levis and Barlaz (2013) concluded that proper waste management (in general) is essential for minimizing risks to human health and the environment. Urban tree waste contains significant amounts of recoverable materials that can be used to produce energy, making waste management a visible and high-impact target for improving environmental sustainability.

In the specific case of the United States, mitigation strategies that affect the electricity mix, the price of energy, and GHG emissions can really change the direction of current waste management (LEVIS; BARLAZ, 2013). Obviously, urban tree waste management systems must have the flexibility to adapt to different compositions, public policies (which are not yet a reality in Brazil), and national electricity production systems, in order to be economically beneficial as well.

1.1.1 Joâo Pessoa and the Clean Development Mechanism (CDM)

Joao Pessoa/PB is in a relatively comfortable situation in terms of air quality, but economic growth trends may cause concern. The inventory of GHG emissions carried out in 2010 by Joao Pessoa's municipal government showed emissions of 1,198,034 t CO_{2-eq}, with an increase of 43.70% for 2012, generating 1,721,681 t CO_2 -eq (JOÂO PESSOA, 2014). For the years 2010 and 2012, the current scenario (landfill) for forestry waste was responsible for 3,520 t CO_2 - eq and 3,810 t CO_{2-eq}, respectively. In 2010, the GHGs emitted corresponded to 0.29% of the total GHGs released in the city, and in 2012, they corresponded to 0.22% of the city's total emissions.

It can be seen that the emission of GHGs generated by tree-planting waste at the landfill corresponds to a small percentage in relation to the total amount released in these years; however, GHGs are cumulative over time, with a global impact, and high levels of concentration can cause damage to the city's existing micro-environment.

According to the Sustainable Joao Pessoa Action Plan (JOÂO PESSOA, 2014), strategies and actions to control GHGs include promoting the use of biomass in the energy matrix and promoting waste recycling. Urban forestry waste is a strong ally in fulfilling this action (JOÂO PESSOA, 2014).

The CDM can be defined as the activities of a project that is part of an enterprise that aims to reduce GHG emissions. It must be exclusively related to certain gases and the sectors/sources of activities responsible for most of the emissions (DUARTE, 2006).

The CDM was established in the Kyoto Protocol as one of the innovation mechanisms, which also includes Joint Implementation (JI) and Emissions Trading (CER). It was an initiative proposed by the Brazilian government during the discussions at the Conferences of the Parties. The proposal consists of the premise that each ton of CO_2 not emitted or removed from the atmosphere by a country can be traded on the world market, creating an attraction for reducing global emissions (SOUZA; RIBEIRO, 2009).

The contribution of CDM projects to landfills in Brazil covers the economic, environmental and social spheres, going beyond the reduction of GHG emissions and becoming a tool for improving MSW services. It has a direct influence on local environmental sustainability, the development of working conditions and the net generation of jobs, as well as attracting and developing technology and regional integration (PAVAN; PARENTE, 2006; CRUZ; PAULINO, 2010).

Brazil stands out for its potential to generate carbon credits in the landfill sector, including the Joao Pessoa metropolitan landfill, which is a promising opportunity to promote municipal development by supporting more appropriate management of solid urban waste (CRUZ; PAULINO, 2010).

In the economic field, such projects contribute to boosting the local economy, since various areas of society are affected by the maintenance, technical assistance and service sectors. It is worth noting that the proceeds from the sale of carbon credits can be divided between the developer and the municipality (PAVAN; PARENTE, 2006).

The carbon emission certificate market can be an important source of funds and an incentive for public authorities to invest in waste management activities without compromising the environmental quality of their surroundings (PAVAN; PARENTE, 2006; DUARTE, 2006). It is worth noting that the last auction for the sale of gas emission certificates on the Sao Paulo stock exchange in 2012 was for €3.30/tCO2 (BM&FBOVESTA, 2012).

Considering the figures for 2012, if Joao Pessoa's metropolitan landfill had a CDM in place, with the sale of a carbon emission certificate from the use of urban forestry waste to generate electricity, the government could have added €13,951.38 to the municipal coffer. In real terms, that would be approximately R$ 36,413.09. Using references from the last trading session of the Sao Paulo stock exchange and the exchange rate at the time.

For a municipality such as Joao Pessoa, which lacks financial sources, setting up a CDM system is a way of raising funds and reducing dependence on the union and the state.

In the environmental field, among the benefits of landfill biogas is the amount of GHG emissions avoided, which is around 80% less, where methane (which accounts for 60% of the gases emitted in landfills and is a gas with a global warming potential 23 times greater than CO_2) accounts for a large proportion of the gases. This can mean more generation of tradable credits when compared to projects that only involve CO_2 reduction (PAVAN; PARENTE, 2006).

The use of biogas appears to be a viable tool for the CDM, since its application for energy purposes can occur immediately, due to its environmentally attractive technology. Since the conversion into energy makes it possible to recover capital and increase the plant's viability (DUARTE, 2006).

According to Rovere et al. (2006), some landfill biogas recovery initiatives underway in the country could generate certified emission reductions (CERs) of 2.3 million t CO_2 avoided/year. With potential revenues of 11.4 million dollars/year, this could be multiplied by 5 with technically feasible initiatives in the short and medium term.

This indicates that the potential for generating carbon credits is very promising, constituting an opportunity to promote social and environmental sustainability by supporting better management of solid urban waste (ROVERE et al., 2006).

With regard to the least impactful scenario in this study (electricity generation), it is worth noting that projects to use solid waste to generate electricity often only consider reducing methane emissions by capturing and burning the gas. Most of the country's landfills (including Joao Pessoa's landfill) rarely recover and burn the methane from the waste, so only capturing and distributing it is already considered a CDM project (ROVERE et al. 2006).

In energy efficiency projects, the use of electricity and the generation of electricity from renewable sources for injection into the grid, the carbon content avoided will determine the amount of reduction certificates issued (CERs) and the revenue from the sale of carbon credits provided by the project (ROVERE et al., 2006).

The first project to implement a CDM system in Brazil was Nova Gerar, in the municipality of Nova Iguaçu/RJ, in 2001. The initial investment consisted of setting up a gas collection system and a modular electricity generating plant (with an expected potential capacity of 12 MW after a few years of operation), to capture methane from the landfill and use it to generate electricity to supply the grid (SOUZA; RIBEIRO, 2009).

According to calculations, the project will prevent the emission of 14.07 million tons of CO_2 over 21 years, by capturing and burning methane from landfill gas to generate electricity. Other benefits include the mitigation of environmental impacts and the generation of social benefits, such as improving the lives of waste pickers and generating employment (SOUZA; RIBEIRO, 2009).

Another project, Vega Bahia, is expected to avoid the emission of 14.5 million tons of CO_2 in the period 2003/2019. With an average annual value of 0.653 million tCO2/year (ROVERE et al., 2006).

The municipality of Sao Paulo, the Secretaria Municipal do Verde e do Meio Ambiente (SVMA), receives 50% of the CERs from the concessionaire company that implemented the CDM to capture gas at the Bandeiras and Sao Joao landfills, as well as a monthly fee for the use of the area and the exploitation of the biogas. As part of the Special Fund for the Environment and Sustainable Development (FEMA), there was a large increase in the budget due to the receipt of CERs. From R$ 32,720,500 in 2007 to R$ 57,366,663.00 in 2008 (CRUZ; PAULINO, 2010).

These funds were used to finance plans, programs and projects in the area of rational use and sustainability of natural resources, control, inspection, defense and recovery of the environment and environmental education. They were used for the Perus Linear Park, the Bamburral Linear Park, the implementation of cycle paths, the veterinary hospital in Anhamquera Park, environmental education projects, selective collection, support for sorting centers and the Perus-Pirapora railroad recovery project, among others (CRUZ; PAULINO, 2010).

In 2007, São Paulo City Hall raised R$34 million from the sale of 808,450 CERs to the Belgian-Dutch bank Fortis, from the Bandeirantes landfill CDM project. Between

2008 and 2009, the municipality received the equivalent of R$71 million, corresponding to 50% of the sale of 1,521,450 CERs. In 2008 alone, the municipality collected €13.689 million (R$37.2 million) from the sale. CERs to the Swiss company Mercuria Energy Trading, whose clients include European oil refineries (RIZZI, 2011).

As happened in the municipality of Sao Paulo, the implementation of a CDM project at the Joao Pessoa metropolitan landfill will bring financial benefits from the sale of CERs. If well directed and managed, these resources should be used for the benefit of society, with actions in the environmental area, improving people's quality of life and social inclusion.

4.4 CONCLUSIONS

From the analysis carried out, it can be concluded that:

The current disposal method used by the municipal authority (landfill) is the least environmentally friendly of all the scenarios studied. There was the highest carbon footprint per ton of urban forestry waste, i.e. emissions of 136.34 kg $_{CO2\text{-}eq/t}$;

The best scenario studied was for the use of tree-planting waste to generate electricity, with a negative environmental impact, and even the heat generation scenario proved to be more environmentally viable than the system used.

The production of electricity is environmentally viable and contributes to the sustainability of the city and can be used as carbon credits;

The implementation of a CDM project at the landfill to make use of urban forestry waste is one of the strategies for managing MSW, with the expectation of promoting great environmental, social and economic benefits, as has happened in the municipalities that have presented such projects.

4.5 REFERENCES

ASSOCIAÇÃO BRASILEIRA DE NORMAS TECNICAS - CATALOGO. **ABNT NBR ISO 14044:** 2009. Corrected version: 2014. 2014a. Available at:

<https://www.abntcatalogo.com.br/norma.aspx?iD=316461>. Accessed on: January 3, 2017.

BRAZILIAN ASSOCIATION OF TECHNICAL RULES - CATaLoGo. **ABNT NBR ISO 14040:**2009. Corrected version: 2014. 2014b. Available at: <https://www.abntcatalogo.com.br/norma.aspx?iD=316461>. Accessed on: January 3, 2017.

ABRELPE. ASSOCIAÇÃO BRASILEIRA DE EMPRESAS DE LIMPEZAS PUBLICAS. **Overview of solid waste in Brazil:** 2014. Sao Paulo. 2015. Available at: <http://www.abrelpe.org.br/panorama/panorama2014.pdf>. Accessed on: April 13, 2017. 2017.

AVALIAÇÂO DO CICLO DE VIDA BRASIL - ACV BRASIL. **Life Cycle Assessment**. Available at: <http://acvbrasil.com.br/avaliacao-do-ciclo-de-vida/>. Accessed on: January 5, 2017.

BARATTA JUNIOR, A. P.; MAGALHÂES, L. M. S. Use of tree pruning waste from the city of Rio de Janeiro for composting. **Revista de Ciências Agro- Ambientais**, Alta Floresta, v. 8, n.1, 2010.

BM&FBOVESTA. **Carbon credit auctions**. Sao Paulo. 2012. Available at: http://www2.bmfbovespa.com.br/Consulta-Leiloes/leiloes-de-credito-de-carbono-login.aspx?idioma=en-br. Accessed on: April 23, 2017.

CAMiLo, D. R.; EsPADA, A. L. V.; MARTiNs, J. R.F. **Characterization of the Management Systems for Pruning and Removal of Urban Tree Waste in the Municipalities of the State of São Paulo**. Piracicaba. 2008. Supervised internship report. Luiz de Quairoz College of Agriculture, University of São Paulo, 2008.

CANADA. information Center Government of Alberta. **Recommendations for Reducing Leaf and Yard in Aberta**. Alberta Environment and Sustainable Resource Development, 2014. Available at: <http://aep.alberta.ca/waste/documents/ReducingLeafYardWaste-Feb2014.pdf>. Accessed on: Feb. 10, 2017.

CARVALHO, M.; FREIRE, R. S.; MAGNO, A. H. .**Promotion of sustainability by quantifying and reducing the carbon footprint: new practices for organizations.** In: Global Conference on Global Warming, 2015, Athens, Greece. Proceedings of the global conference on global warming, 2015.

MINAS GERAIS ENERGY COMPANY. **Manual de arborizaçao**. Belo Horizonte: MG, p. 112, 2011. Available at:

<http://www.cemig.com.br/sites/imprensa/pt-br/Documents/Manual_Arborizacao_Cemig_Biodiversitas.pdf>. Accessed on: April 10, 2017.

PARANAENSE ENERGY COMPANY. **Arborization of public roads**. Curitiba: PR, 2009. Available at:

<http://www.copel.com/hpcopel/guia_arb/copel_e_a_arborizacao_de_vias_publicas.html>. Accessed on: January 10, 2017.

CORTEZ, C. **Study of the potential use of biomass resulting from urban tree pruning for energy generation:** case study: AES Eletropaulo. 2011. Thesis (Doctor of Science). University of Sao Paulo, Sao Paulo, SP, 2011.

CRUZ, S. R. S.; PAULINO, S. R.; Clean development mechanism (CDM) project for landfills and solid waste management in the city of Sao Paulo. In: V Encontro Nacional da Anppas, 2010, Florianópolis. **Anais ...**, V Encontro Nacional da Anppas, 2010. Available at: <http://www.anppas.org.br/encontro5/cd/artigos/GT3-7- 389-20100903200350.pdf>. Accessed on: April 25, 2017.

DUARTE, A. C. **CDM projects in landfills in Brazil**: an alternative for sustainable development. 125 f. 2006. Dissertation (Master's Degree in Water Resources and Environmental Engineering). Federal University of Paranâ. Curitiba, 2006. Available at: <http://www.ppgerha.ufpr.br/publicacoes/dissertacoes/files/127- Adriana_Carneiro_Duarte.pdf>. Accessed on: January 23, 2017.

ECOINVENT CENTER. **What we do**. Zurich, 2015. Available at: <http://www.ecoinvent.org/about/about.html>. Accessed on: January 16, 2017.

EMLUR. SPECIAL MUNICIPAL CLEANING AUTHORITY FOR JOÂO PESSOA. **Information on annual pruning and costs**. Joao Pessoa. 2016.

EMLUR. SPECIAL MUNICIPAL AUTHORITY FOR URBAN CLEANING IN JOÂO

PESSOA. **Plano Municipal de Gestao Integrada de Residuos Sólidos-PMGIRS**: Diagnosis and Planning of Urban Cleaning and Solid Waste Management Services. Joao Pessoa. v. 2, 2014. Available at:<http://transparencia.joaopessoa.pb.gov.br/dadospublicos/?p=111>. Accessed on: January 13, 2017.

ENVIRONMENTAL PROTECTION AGENCY. **Food Wast**. 2015. Available at: <http://www3.epa.gov/epawaste/conserve/tools/warm/pdfs/Food_Waste.pdf>. Accessed on: February 10, 2017.

FERREIRA, J. V. R. **Product Life Cycle Analysis**. Viseu: Polytechnic Institute of Viseu, 2004.

FREIRE, R. S.; CARVALHO, M.; CARMONA, C. U. M.; MAGNO, A. H. **Perspectives on the implementation of climate change public policies in Brazil.** In: Global Conference on Global Warming, 2015, Athens, Greece. Proceedings of the global conference on global warming, 2015.

GALDIANO, G. P. **Life Cycle Inventory of Offset Paper Produced in Brazil**.

2006. Dissertation (Master's in Engineering). University of Sao Paulo, Sao Paulo, 2006.

GOLVEIA, N. Residuos sólidos urbanos: impactos Socioambientais e perspectiva de manejo sustentâvel com inclusao social. **Ciência & Saùde Coletiva**, v. 17, n. 6, p. 1503:1510, 2012.

GUINÉE JB. (ed) **Life Cycle Assessment: An operation al guide to the ISO Standards;**

LCA in Perspective; Guide; Operational Annexto Guide. Center for Environmental Science, Leiden University, The Netherlands, 2001.

GUINEA JB. **Handbook on life cycle assessment:** operation al guide to the ISO standards. Kluwer Academic Publishers, Boston, 2002.

INTERGOVERNMENTAL PANEL ON CLIMATE CHANGE. r **Who is the IPCC?** Geneva: Secretariat of the WMO, 2013. Available at: <http://www.ipcc.ch/>. Accessed on: January 5, 2017.

INTERGOVERNMENTAL PANEL ON CLIMATE CHANGE. **History**. Geneva: Secretari at of the WMO, 2015. Available at:

<http://www.ipcc.ch/organization/organization_history.shtml>. Accessed on: April 1, 2017.

JOÂO PESSOA. Joao Pessoa City Hall. Joao **Pessoa Sustainable Action Plan 2014.** Joao Pessoa, 2014. Available at:

<http://www.joaopessoa.pb.gov.br/>. Accessed on: January 3, 2017.

KLOPFFER, W. **The critical review of life cycle assessment studies according to ISO 14040 and 14044:**origin, purpose and practical performance. The International Journal Of Life Cycle Assessment, Heidelberg, p. 1-7. 2012.

LEVIS, J. W.; BARLAZ, M. A. **Composting Process Model Documentation**; Project Report, North Carolina State University: Raleigh, NC. 2013. Available at: <http://www4.ncsu.edu/~jwlevis/Composting.pdf>. Accessed Feb. 11, 2017.

MARTINS, C. H. The use of wood from the pruning of road trees in Maringa/PR. **Revista Verde**, v. 8, n. 2, p. 257 - 267, 2013.

MEIRA, A. M. **Urban tree waste management**. 2010. 179 f. Thesis (Doctorate in Science). Luiz de Queiroz College of Agriculture, University of São Paulo, Piracicaba, 2010.

MORRIS, J.; MATTHEWS, H. S.; MORAWSKI, C. **Review of LCAs on Organics Management Methods & Development of an Environmental Hierarchy**. nformation Center Alberta Enviroment. Alberta, 2011. Available at:<http://environment.gov.ab.ca/info/library/8350.pdf>. Accessed on: Feb. 10, 2017.

PAVAN, M. C. O.; PARENTE, V. CDM projects in Brazilian landfills: political, socioeconomic and environmental analysis. In: XXX Inter-American Congress of Sanitary and Environmental Engineering, 2006, Punta del Este. Proceedings ... XXX Inter-American Congress of Sanitary and Environmental Engineering, 2006. Available at: <http://www.cetesb.sp.gov.br/wp-content/uploads/sites/27/2014/01/parente_pavan.pdf.> Accessed on: April 23, 2017.

PIRES, A.C; RABELO, R. R; XAVIER, J. H. V. **Potential use of Life Cycle Analysis (LCA) associated with organic production concepts applied to family farming**. Cadernos de Ciência e Tecnologia, Brasilia, v. 19, n. 2, p. 149-178, 2002.

RIZZI, C. A. A **questão da participação da comunidade do Distrito de Perus (Sâo Paulo/Brasil), no projeto MDL Aterro Bandeirantes, Confins** [Enligne], 2011. Available at:

<https://confins.revues.org/6870?lang=pthttp://www.revistas.sp.senac.br/index.php/ITF/article/viewFile/97/122>. Accessed on: April 25, 2017.

SOUZA, G. D.; RIBEIRO, W. C. Nova Gerar: Brazil's pioneering CDM experience. **Cronos**, v. 10, n. 2, p. 15-34, 2009.

NATIONAL SECRETARIAT FOR ENVIRONMENTAL SANITATION. National Sanitation Information System. **Diagnosis of urban solid waste management - 2014.** Part 2-Table of Information and Indicators. Brasilia: MCIDADES. SNSA, 2016

SIMAPRO. **SimaProDatabase8**:Manual Methods Library. California, 2015. Available at: <https://www.pre-sustainability.com/simapro-database-and-methods-library>. Accessed on: April 4, 2017. 2017.

RECIFE. Department of the Environment and Sustainability. **Manual de arborização:** orientação e manutençâo da arborização urbana da cidade do Recife. Ed. Recife: [S.N.]. 2013. Available at:<www2.recife.pe.gov.br/wp- content/uploads/Manual_Arborizacao.pdf>. Accessed on: April 13, 2017.

ROVERE, E. L. La; COSTA, C. V.; DUBEUX, C. B. S. **Landfill in Brazil and the Clean Development Mechanism (CDM)**: Opportunity to promote socio-environmental development. Web-Resol, Rio Bonito, 2006. Available at: <http://www.web-resol.org/textos/28-La%20Rovere%20E.pdf.> Accessed on: April 22, 2017.

VALINHOS. Valinhos City Hall. **Integrated Solid Waste Management Plan for the Municipality of Valinhos-SP**. Valinhos, 2011. Available at: <http://www.valinhos.sp.gov.br/portal/arquivos/planejamento/PGIRS_- _Verso_Preliminar_2.pdf>. Accessed on: February 10, 2017.

5. ARTICLE 4 - TEMPORAL ANALYSIS OF THE VOLUME OF WASTE FROM URBAN ARBORIZATION IN JOAO PESSOA

LUIZ MOREIRA COELHO JUNIOR; YURI ROMMEL VIEIRA ARAÙJO; ZAYNE CHRISTINA GONÇALVES MOEIRA; THIAGO FREIRE MELQUiADES

SUMMARY

Urban tree planting requires management actions to ensure sustainability in the city. The generation of urban tree waste is recurrent to the need for maintenance over time. This study analyzed the time series of woody volume from urban tree planting in the city of Joao Pessoa/PB, from January 2008 to December 2014, in order to characterize the volume of urban tree planting waste and adjust a forecasting model. The models studied were from the ARIMA (*Autoregressive Integrated Moving Average*) family. The main results obtained were: The ARIMA family models showed satisfactory results for forecasting; the ARIMA (0,1,4) model was the one that provided the best forecast for the year 2014; this study provides a better understanding of the volume behavior of urban tree waste in Joao Pessoa and, which helps guide municipal public policies.

Keywords: forest economy, biomass, ARIMA.

5.1 INTRODUCTION

Urban tree planting requires management actions to ensure sustainability in the city. This management results in the generation of a significant amount of biomass waste, which is classified as solid urban waste (MSW) (BRASIL, 2007, 2010; RECIFE, 2013).

In 2014, Brazil generated 78.6 million tons (t) of solid urban waste (MSW), including residential waste and urban cleaning (sweeping, weeding and pruning). In 2014, Paraiba generated 1.278 million tons of MSW, an average of 3,504 tons/day (ABRELPE, 2015).

According to the National System of Information on Environmental Sanitation (SNISA), in the municipalities of Belo Horizonte (MG), Goiana (PE), Jaboatao dos

Guararapes (PE), and Igaracy (PB), urban tree waste accounted for 9.37%, 1.41%, 1. 48% and 7.10%,

respectively, of MSW destined for processing plants in 2014 (SNSA, 2016).

In the municipality of Joao Pessoa, 415.96 thousand tons of MSW were generated in 2013, including household waste, rubble and urban cleaning waste. Of this total, 28.71 thousand tons came from urban tree maintenance and tree replacement, corresponding to 6.9% of MSW (EMLUR, 2016).

The generation of urban tree waste is recurrent throughout the year, due to the need to maintain the vegetation, where this periodicity over time is classified as a time series. A time series is defined as a set of observations of a variable, arranged sequentially over time (ANDRADE, 2013).

Time series analysis makes it possible to infer past behavior and generate information on probable future behavior by building models that predict future movements (FISCHER, 1982).

The use of time series forecasting models is one of the possible alternatives in the decision-making process involving activities that require planning, political evaluation and the reduction of uncertainty. It has various applications, with different resources and fields of knowledge, such as the administrative area, economics, the forestry sector, the health area, among others (ANTUNES; CARDOSO, 2015; BRESSAN, 2004).

In the forestry sector, there are several applications of time series analysis. The following studies stand out: Floriano et al. (2006) to develop height growth equations for a population of *Pinus elliottii*; Coelho Junior et al. (2006) to predict the price of a meter of charcoal in the state of Minas Gerais; Soares et al. (2010), who developed a model to predict the price of standing timber of *Eucalyptus spp*, in Itapeva (SP) and Bauru (SP); Coelho Junior et al. (2009) forecasting natural rubber prices on the international market; Soares et al. (2008) forecasting natural rubber prices on the national market, and; Almeida et al. (2009) forecasting the price paid for plywood exports from Paraná, where graphical and statistical analysis indicated the ARIMA (1,1,3) model as the best fit for the plywood price series.

However, there are no studies that show the predicted volume of biomass from urban tree maintenance, which is extremely important for decision-making in urban tree planning. This study analyzed the predicted volume of wood from urban tree planting in the city of Joao Pessoa/PB.

5.2 MATERIALS AND METHODS

5.2.1Object of study

The study used the historical series of urban tree maintenance waste in tons (t) from Joao Pessoa, collected monthly by the Urban Cleaning Autarchy (EMLUR). The period analyzed was from January 2008 to December 2014, which contains 84 pieces of information. Data from January 2008 to December 2013 was used to adjust the model. The period from January to December 2014 was used to validate the forecasting model.

5.2.2Time series analysis

A time series $\{Y_i, t = 1,2,\text{-},n\}$ is defined as a set of observations of a variable arranged sequentially over time (MORETTIN; TOLOI, 2006). ARIMA (Autoregressive Integrated Moving Average) models, introduced by Box and Jenkins (1976), are based on the idea that a non-stationary, homogeneous time series can be modeled using "d" differentiations and the inclusion of an autoregressive component "p" and a moving average component "q". Since $\{Y_t\}$ is a process, it can be described using an *ARIMA (p, d, q)* model as follows:

$$\phi_p(B)Z_t = \theta_0 + \theta_q(B)a_t$$

where,

$$Z_t = \begin{cases} Y_t\text{, if the process is stationary, when} = 0 \\ (1 - B)^d\, Y_t\text{, if the process is not stationary, when d} \geq 1 \end{cases}$$

The weighting of the differentiation of Y_t corresponds to an ARIMA model (p, d, q) with:

$$\phi_p(B)(1-B)^d Y_t = \theta_0 + \theta_q(B)a_t$$

where $\phi_p(B)=1-\phi_1 B-\phi_2 B^2-\ldots-\phi_p B^p$ is the p-order autoregressive operator $[AR(p)]$, $\theta_0=\mu(1-\phi_1-\phi_2-\ldots-\phi_p)$ is the intercept or constant, $\theta_0(B)=1-\theta_1 B-\theta_2 B^2-\ldots-\theta_q B^q$ is the q-order moving average operator $[MA(q)]$ and is a *white* noise process. If the constant θ_0 is non-zero, the integrated series will have a deterministic trend, i.e. the series will have an upward or downward trend, which is independent of random disturbances (PINDYCK; RUBENFIELD, 1991).

To verify the stationarity of the model, visual analysis and decomposition of the time series were carried out. To verify the stationarity of the series, we used the Augmented Dickey-Fuller (ADF) test, developed by Dickey and Fuller (1981), the Phillips-Perron (PP) test; developed by Phillips and Perron (1988) and the Kwiatkowski-Phillips-Schmidt-Shin (KPSS) test; developed by Kwiatkowski and others (1992). These tests confirm whether or not the *Y* series$_t$ has a unit root, and thus whether or not it is stationary.

Identifying the model consists of determining its order based on the "principle of parsimony". This stage is the most critical phase of model use and consists of determining the types of model generating the series, which are called:

	AR		(p)
	MA		(q)
	ARMA	What is the order of the model?	(p' q)
y_t	ARIMA	i.e. what are the values of	(p,d,q)
	SARIMA		
	.		(p,d,q) x (P,D,Q)s
	.		
	.		

Time-domain analysis (BOX and JENKINS, 1976) was used to help in this identification stage, as this is the fundamental approach to time series analysis.

After identifying and choosing the appropriate model, the parameters of the AR process and the parameters of the MA process were estimated. The parameters were estimated using the Gaussian distribution by the Maximum Likelihood method, for all possible combinations, in order to satisfy the conditions of invertibility and uniqueness of the parameters.

Next, to check whether the proposed model is white noise, diagnostics were carried out using standardized residual analysis, residuals of the autocorrelation function (ACF), residuals of the partial autocorrelation function (PACF) and the portmanteau test of Box and Pierce (1970):

$$Q_k = n\sum_{1}^{k} c_k^2$$

where n = number of observations; k = number of lags; and ck = autocorrelation of the residuals. The model is accepted if $Q \leq x^2\ (\lambda, k\text{-N})$, where x^2 is the chi-square, λ is the significance level (with a 95% confidence interval), k is the lag order and N is the number of parameters.

Another way of checking the model is using the Akaike Information Criterion (AIC), $AIC = {}^{-}2\ln(L)+2(p+q)$ where L = maximum likelihood; p and q = model parameters, in order to obtain the minimum AIC value (AKAIKE, 1977).

Once the iterative process of identifying, estimating and checking the model has been completed, if it provides an estimate of the series that satisfactorily fits the actual data, then it can be used to predict the values of the variable.

Forecasting processes, which use time series models, are procedures that aim to extend the model described and adjusted to the present and past values of the variable to future values. Therefore, forecasting makes it possible to determine the expected value of a future observation.

The Mean Squared Error (MSE) of the forecasts obtained was checked, allowing a comparison of the predicted and observed values of the adjusted series, making it possible to choose the model with the lowest MSE (MORETTIN; TOLOI, 2006).

$$EQM = \frac{\sum (y_t - y_t^e)^2}{n}$$

5.3 Results and discussion

Figure 5.1 shows the evolution and behavior of the Joao Pessoa urban forestry waste mass series (MRAU-JP) and the logarithmized series [Ln(MPA-JP)], from January 2008 to December 2013, expressed in 1,000 tons. Neperian logarithmization was necessary to stabilize the variance while preserving the properties of the series data. The MRAU-JP has an estimated mean of 2,151, a median of 2,115, a minimum value of 1,256 and a maximum value of 2,772.

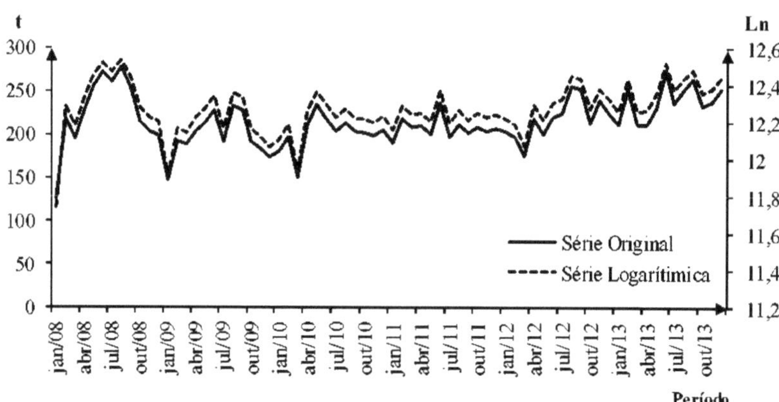

Figure 5.1. Behavior of Joao Pessoa's Urban Forestry Waste Mass Series, original and logarithmic from 2008 to 2013 in 1,000 (t).

We can see an upward trend in MRAU-JP over time, which indicates that it is non-stationary. Another behavior observed is the increase in the mass collected from April to July, compared to the other months of the year. This behavior is due to the period of highest rainfall in the city, when there is more damage to the trees, with branches breaking and trees falling.

It is also worth noting that when analyzing the mass of biomass generated each year, one must take into account the demand coming mainly from the population for the service, by requesting the competent body to authorize and carry out maintenance and suppression. It should also be noted that the waste management policy implemented by the government also influences the amount of waste generated each year, such as the hiring or firing of labor, changes in administrative managers, revisions and alterations to contracts with service providers, among others.

Figure 5.2 shows the decomposition of the data, trend, seasonality and residuals of Ln (MRAU-JP), for the period from January 2008 to December 2013. The decomposition of the series shows the presence of the trend and seasonality components, which should be inferred in the model. Followed by the behavior of the residuals. The largest characteristics are shown by the size of the gray bars: the smaller the bars, the more predominant the component; the larger the bars, the less predominant the component in the MRAU-JP decomposition.

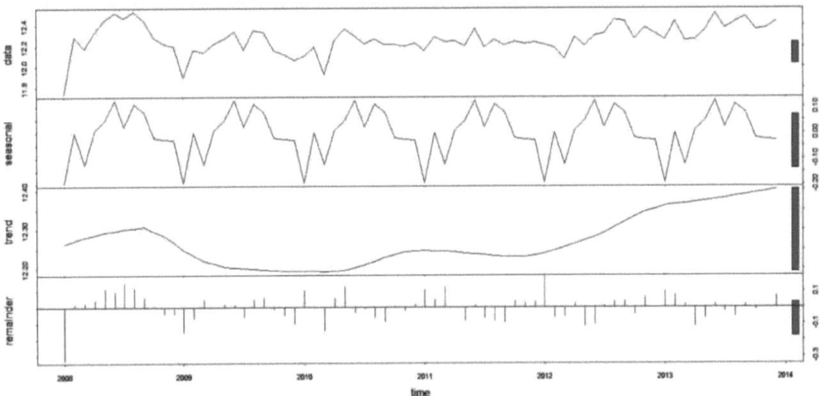

Figure 5.2. Data breakdown, trend, seasonality and residuals of the series of logarithmized urban tree waste mass in Joao Pessoa, from Jan/2008 to Dec/2013.

Visual verification and the decomposition of the series alone cannot tell whether the series [Ln(MRAU-JP)] is stationary or not. So, for a more formal verification, we used the Augmented *Dickey-Fuller* (ADF) test, the *Phillips-Perron* (PP) test and the *Kwiatkowski-Phillips-Schmidt-Shin* (KPSS) test to check for the presence of a unit root, as shown in Table 5.1.

The ADF test shows the presence of stationarity when the null hypothesis (H_0) is rejected, i.e. when the series has a unit root. And for the alternative hypothesis (H_1), the stationarity of the series is confirmed, without the presence of a unit root. The ADF test shows that if $|\alpha| < |t|$, H_0 is accepted. Thus, Ln (MRAU-JP) at a significance level of 5% accepts H_0 , because there is a unit root, i.e. it is non-stationary, and it is necessary to transform it by the 1ª difference to make it stationary.

Table 5.1. Unit root tests for the logarithmized Joao Pessoa urban afforestation residual series [Ln(MRAU-JP)] and its first difference {ia Dif [Ln(MRAU-JP)]}.

Tests	[Ln(MRAU-JP)]				1ª Dif [Ln(MRAU-JP)]			
	1%	5%	10%	t	1%	5%	10%	t
ADF	4,04	-3,45	-3,15	-3,41	-4,04	-3,45	-3,15	-8,08
KPSS Test	0,73	0,46	0,34	0,40	0,73	0,46	0,34	0,09
Phillips-Perron	-4,09	-3,47	-3,16	-6,29	-4,09	-3,47	-3,16	-13.98

Applying the ADF test to the 1st difference of the 1st Dif [Ln(MRAU-JP)] series, it was observed that for the significance levels studied, the t value was greater than any of the critical values, so it can be said that there is no unit root, thus stating that the F Dif [Ln(MRAU-JP)] is stationary. The KPSS test states that the null hypothesis is that there is no unit root, so [Ln(MRAU-JP)] showed t = 0.40, which is greater than the test's critical value of 10%. This shows the rejection of the null hypothesis, showing that [Ln(MRAU-JP)] is non-stationary. For the F Dif [Ln(MRAU-JP)], t = 0.09, lower than any of the critical values, characterizing a possible stationarity of the series in first difference. The PP test for the 1st Dif [Ln(MRAU-JP)] proves stationarity, showing a t = -13.98, well above the t = -6.29 of the MRAU-JP at level (table 2.1).

Figure 5.3 shows the identification of the model order using the Ln(MRAU-JP) Autocorrelation Function (ACF) and Partial Autocorrelation Function (PACF). It can be seen that the ACF graph decays rapidly after lag 4, which indicates that the series cannot be stationary and does not show seasonal behavior over time. This means that the first order of differentiation is necessary for the series to be stationary. Considering that there is no cutoff after lag1, this suggests an MA (0). As for the PACF graph, only lag 1 and 2 are significant, which indicates an autoregressive term of order AR (2). This identified the ARIMA (2,1,0) model. However, according to Meyler (1998) the interpretation of the Autocorrelation (ACF) and Partial Autocorrelation (PACF) graphs can be difficult and the identification of models using the Box-Jenkins methodology involves a certain amount of subjectivity.

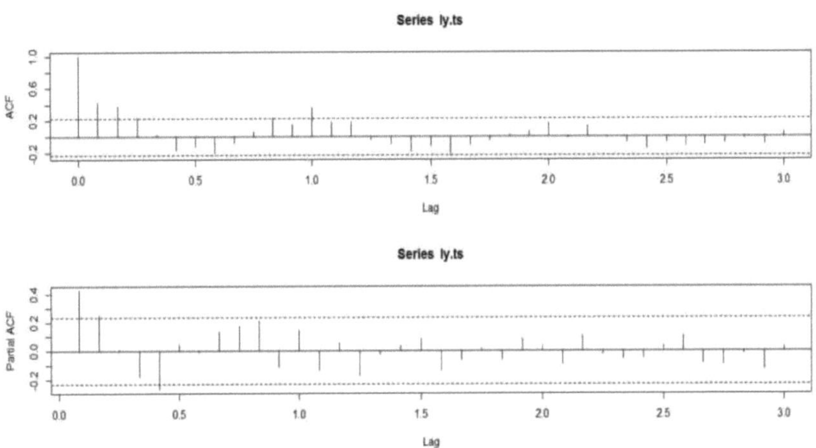

Figure 5.3. Autocorrelation and partial autocorrelation of Ln(MRAU-JP) in level.

However, once the ARIMA model (2,1,0) had been identified, the order was set at a maximum of 5 lags for the autoregressive processes [AR (p = 5)] and 5 lags for the moving average processes [MA (q = 5)]. A sample space of 33 ARIMA models (p,d,q) was constituted.

Of the 33 models, only 24 showed white noise, verified using the Box-Pierce test, where $Q(m) < x^{x}_{a}$. Checking the residual values graphically using the *tsdiag (x)* function in R, only 12 of the 24 models were detected as white noise and were separated for forecasting. The ARIMA (2,1,0) model initially selected did not meet the criteria adopted.

Of the 12 models separated after checking for the presence of white noise, only four showed significant values at a significance level of 80% and 95%. Table 5.2 shows the results presented by the models for the AIC and Box-Pierce test.

Table 5.2. Models selected for predicting the time series of mass from urban trees in Joao Pessoa, from January to December 2014.

ARIMA (p,d,q)	AIC	Test Box-Pierce	
		Q $^{(m)}$	*Xa*
(5,1,0)	-90,317	0,2247	40,979
(1,1,2)	-91,072	0,1802	44,687
(0,1,3)	-91,92	0,1462	46,05
(0,1,4)	-89,951	0,1048	46,938

With this analysis, the selected models were predicted and the Mean Square Error of Prediction (MSEE) was used to determine which was the most appropriate. Table 5.3 shows the forecasts for the biomass volume of trees in Joao Pessoa from January to December 2014. It can be seen that the ARIMA model (0,1,4) was the one with the lowest EQMP for the year 2014, making it the model of choice. The ARIMA model (0,1,4) for 12 periods chosen was:

$$Y_t = \frac{\left(1+\theta_1 B+\theta_2 B^2+\theta_3 B^3+\theta_4 B^4\right)\alpha_t}{(1-B)}$$

$$Y_t = \frac{\left(1-0{,}4750+0{,}2984+0{,}209-0{,}0265\right)\alpha_t}{(1-B)}$$

Figure 5.4 shows the residual values of the ARIMA model (0,1,4). In addition to the Box-Pierce test, the correlogram suggests independence of the residuals for various lags, where the control limits of the CAF graph corroborate that the model chosen was adequate.

Table 5.3. Observed values and predictions by ARIMA models (p,d,q) for 12 periods in 2014 for the mass of urban tree waste in Joao Pessoa (1,000 t).

Period	Observed	(5,1,0)	(1,1,2)	(0,1,3)	(0,1,4)
Jan/14	223,75	237,69	236,43	236,28	236,48
Feb/14	260,28	240,41	242,44	241,37	242,22

Mar/14	214,33	250,64	244,36	245,71	246,77
Apr/14	248,32	243,00	244,97	245,71	246,21
May/14	279,88	241,64	245,16	245,71	246,21
Jun/14	327,76	248,53	245,22	245,71	246,21
Jul/14	255,38	243,98	245,24	245,71	246,21
Aug/14	277,62	242,50	245,24	245,71	246,21
Sep/14	248,62	246,35	245,24	245,71	246,21
Oct/14	238,39	244,63	245,24	245,71	246,21
Nov/14	257,96	242,98	245,24	245,71	246,21
Dec/14	170,09	245,53	245,24	245,71	246,21
EQMP		4,62	4,36	3,81	3,1
Error (%)		-2,5	-2,4	-2,25	-2,03

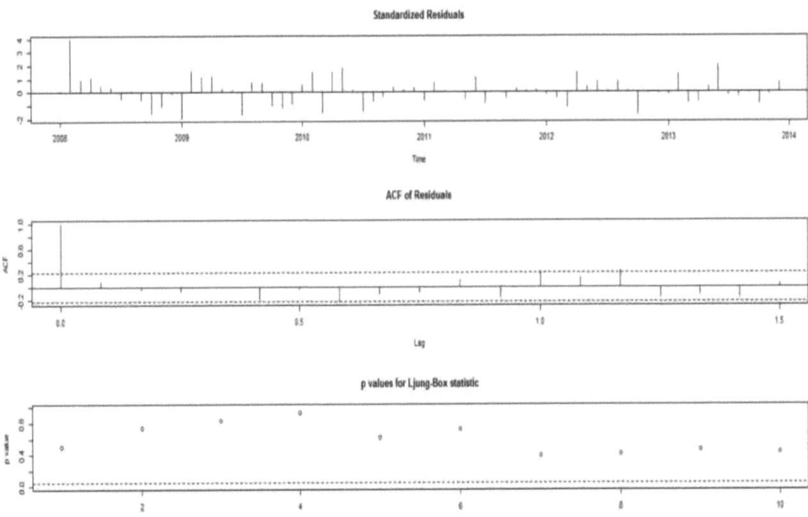

Figure 5.4 Residuals from the Ln(MRAU-JP) ARIMA (0,1,4) model.

It was decided to project values for 12 months for the year 2014. Figure 5.5 shows the behavior of the forecast, considering a confidence interval of 80% and 95%, as well as the observed values for 2014 in months. The model presented satisfactory forecasts, with an EQMP of 3.10.

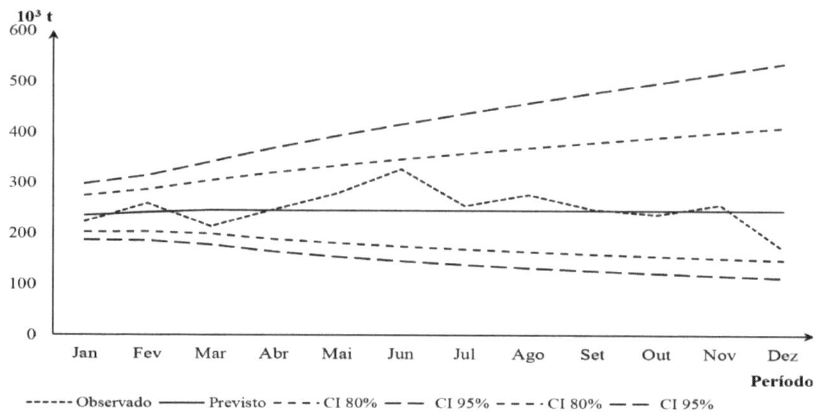

Figure 5.5 2014 forecast of tree maintenance mass in Joao Pessoa (1,000 t).

5.4 CONCLUSIONS

For the conditions developed in this work, it is concluded that:

The models in the ARIMA family provided satisfactory forecasting results. The ARIMA (2,1,0) model identified through the ACF and PACF did not present adequate adjustments, making it necessary to look for new models.

For the sample space of the 33 models chosen for 5 lags for the moving average processes and 5 for the autoregressive component, the ARIMA (0,1,4) model was the one that provided the best forecast for the 12 periods of the volume of mass generated in tree maintenance in Joao Pessoa, with the lowest EQMP.

This study provides a better understanding of the volume behavior of urban tree waste in Joao Pessoa, with a view to helping guide municipal public policies.

5.5 REFERENCES

AKAIKE, H. On entropy maximization principle. In: KRISHAIAH, P. R. (Ed.). **Application of statistics**. Amsterdam, The Netherlands: North-Holland, 1977. 27-41 p.

ALMEIDA, A. N.; SOUZA, V. S.; LOYOLA, C. E.; BITTENCOURT, M. V. L.; SILVA, J. C. G. L. Analysis of the external price of plywood from Paraná using the Box & Jenkins methodology. **Scientia Forestalis**, v. 37, n. 81, p. 061-069, 2009.

ANDRADE, B. S. **Statistical Approach in Models for Counting Time Series**. 2013, 146 f

Dissertation (Master in Statistics), Federal University of Sao Carlos, Sao Carlos, 2013.

ANTUNES, J. L. F.; CARDOSO, M. R. A. Use of time series analysis in epidemiological studies. **Epidemiologia e Serviço de Saùde**, 24 (3), 565-576, 2015.

ABRELPE - BRAZILIAN ASSOCIATION OF PUBLIC CLEANING AND SPECIAL WASTE COMPANIES. **Overview of solid waste in Brazil: 2014**. Sao Paulo: ABRELPE. Available at: <http://www.abrelpe.org.br/panorama_edicoes.cfm>. Accessed on: March 1, 2017.

EMLUR - SPECIAL MUNICIPAL CLEANING AUTHORITY FOR JOAO PESSOA. **Information on annual pruning and costs**. Joao Pessoa. 2016.

BOX, G. E. P.; JENKINS, G. M. **Time series analysis: forecasting and control**. San Francisco: Holden-Day, 1976.

BRAZIL. Law No. 11.445, of January 5, 2007. **Establishes national guidelines for basic sanitation and other measures**. Available at:

<http://www.planalto.gov.br/ccivil_03/_Ato2007-2010/2007/Lei/L11445.htm#art7>. Accessed on: March 18, 2017.

. Law No. 12.305, of August 2, 2010. **Establishes the National**

Solid Waste and other measures. Available at:

<http://www.planalto.gov.br/ccivil_03/_ato2007-2010/2010/lei/l12305.htm>. Accessed on: March 15, 2017.

BRESSAN, A. A. Decision-making in agricultural futures as time series forecasting models. **Revista de Administraçâo de Empresas**, v. 3, n. 1, 2004.

COELHO JUNIOR, L. M.; REZENDE, J. L. P.; CALEGARIO, N.; SILVA, M. L. Longitudinal analysis of charcoal prices in the state of Minas Gerais. **Revista Arvore**, v. 30, n. 3, p. 429-438, 2006.

COELHO JUNIOR, L. M.; REZENDE, J. L. P.; SAFADI, T.; CALEGARIO, N.; Analysis of the temporal behavior of rubber prices in the international market. **Ciência Florestal**, v. 19, n. 3, 2009.

FISCHER, S. **Univariant time series: Box and Jenkins methodology**. Porto Alegre: Economics and Statistics Foundation, 1982. 186 p. Available at: <https://cdn.fee.tche.br/teses/digitalizacao/teses_4.pdf>. Accessed on: April 1, 2017.

FLORIANO, E. P; MULLER, I.; FINGER, C. A. G.; SCHNEIDER, P. R.; Adjustment of traditional models for time series of tree height data. **Ciência Florestal**, v. 16, n. 2, 2006.

MEYLER, A.; KENNY, G.; QUINN, T. **Forecasting Irish inflation using ARIMA models**. 1998.

46 p. Available at:
<http://www.centralbank.ie/publications/documents/3RT98.pdf>. Accessed on April 5, 2017.

MORETTIN, P. A.; TOLOI, C. M. C. **Anàlise de séries temporais**. 2. ed. ver. e ampl. Sao Paulo: Edgar Blücher, 2006.

PINDYCK, R. S.; RUBENFIELD, D. L. **Econometric models and economic forecasts**. 3rd ed. New York: McGrawHill, 1991.

RECIFE. Secretariat for the Environment and Sustainability (SMAS). **Manual de arborizaçâo: orientações e procedimentos técnicos básicos para a implantaçâo e manutençâo da arborizaçâo da cidade do Recife**. Ed. Recife: [s.n.]. 2013.

SNSA - NATIONAL SECRETARIAT FOR ENVIRONMENTAL SANITATION. National Sanitation Information System. **Diagnosis of urban solid waste management-2014.** Part 2 - Table of Information and Indicators. Brasilia: MCIDADES. SNSA, 2016.

SOARES, N. S.; SILVA, M. L.; LIMA, J. E.; CORDEIRO, S. A. Forecast analysis of the price of natural rubber in Brazil. **Scientia Forestalis**, v. 36, n. 80, p. 285-294, 2008.

SOARES, N. S.; SILVA, M. L.; REZENDE, J. L. P.; LIMA, J. E.; CARVALHO, K. H. A.; Elaboration of a model for predicting the price of *Eucalyptus spp*. wood **Cerne**, v. 16, n. 1, p. 41-52, 2010.

VENABLE, W. N.; RIPLEY, B. D. **Modern applied statistic with S-PLUS**. 3. ed. New York: Springer-Verlag, 1999. 501 p.

6. FINAL CONSIDERATIONS

After the analysis carried out in this dissertation, the following final considerations were reached:

In general, this study has provided significant information to guide public policy decisions on the management of urban tree waste.

By means of sampling, it presented a diagnosis of the current situation of Joao Pessoa's urban tree planting and suggested some ways of conducting tree planning in the city. Therefore, more in-depth studies and a sensu are needed for better planning of urban tree planting.

The two analyses of the environmental impacts caused by the disposal of urban forestry waste show that the current form is the most damaging to the environment. The first looked at the carbon footprint in four waste disposal scenarios and the second looked at the energy use strategy as an instrument of the Clean Development Mechanism (CDM). Therefore, this analysis shows that the way in which urban forestry waste is disposed of in Joâo Pessoa needs to be rethought. Economic studies should be carried out showing alternative ways of using this available biomass for energy.

And by analyzing the historical series of urban tree maintenance waste generation, he presented a model for the generation of this waste.

Finally, laboratory tests on the calorific value of the species that make up Joâo Pessoa's urban forest should be carried out to help determine the feasibility of using biomass for energy. This could be tested with

different technologies and methods of conversion: use *in natura; in the* form of briquettes; by the gasification technique, and; by the anaerobic process, etc.

7. REFERENCES

ABRELPE. BRAZILIAN ASSOCIATION OF PUBLIC CLEANING and special waste COMPANIES. **Overview of solid waste in Brazil:** 2014. Sao Paulo.

2015. Available at: <http://www.abrelpe.org.br/Panorama/panorama2014.pdf>. Accessed on: November 13, 2016.

ABNT. BRAZILIAN ASSOCIATION OF TECHNICAL STANDARDS - CATALOG.

ABNT NBR ISO 14044: 2009. Corrected version: 2014. 2014 A. Available at:

<https://www.abntcatalogo.com.br/norma.aspx?ID=316461>. Accessed on: April 3, 2017. 2017.

. **NBR ISO 10004: Solid Waste: Classification**. Rio de Janeiro, 2004. 77 p.

. **ABNT NBR ISO 14040: 2009** Corrected Version: 2014**.** 2014b.

Available at: <https://www.abntcatalogo.com.br/norma.aspx?ID=316461>. Accessed on: April 3, 2017. 2017.

AGUIRRE, L. A.. **Introduction to system identification:** linear and non-linear techniques applied to real systems. UFMG, 2007. ISBN 85-7041-220-7. 3rd Revised and Expanded Edition. 2007.

AKAIKE, H. On entropy maximization principle. In: KRISHAIAH, P. R. (Ed.). **Application of statistics**. Amsterdam, The Netherlands: North-Holland, 1977. 27-41 p.

ALBERTIN, R. M.; De ANGELIS F.; De ANGELIS NETO R.; De ANGELIS B. L. D. Quali-quantitative diagnosis of the roadside afforestation of Nova Esperança, Paranà, Brazil.

Revista da Sociedade Brasileira de Arborizaçao Urbana, v.6 n. 3, p. 128-148, 2011.

ALMEIDA, A. N.; SOUZA, V. S.; LOYOLA, C. E.; BITTENCOURT, M. V. L.; SILVA, J. C. G. L. Analysis of the external price of plywood from Paraná using the Box & Jenkins methodology. **Scientia Forestalis**, v. 37, n. 81, p. 061-069, 2009.

AMÂNCIO, C. T.; NASCIMENTO, L. F. C. Asthma and environmental pollutants: a time series study. Ver. **Associaçao Médica Brasileira**, 58(3), 302-307 p, 2012.

AVALIAÇÂO DO CICLO DE VIDA BRASIL - ACV BRASIL. **Life Cycle Assessment**. Available at: <http://acvbrasil.com.br/avaliacao-do-ciclo-de-vida/>. Accessed on April 18, 2017.

ANDRADE, B. S. **Statistical Approach in Models for Counting Time Series**. 2013, 146 f Dissertation (Master in Statistics), Federal University of Sao Carlos, Sao Carlos, 2013.

ANGELIS, B. L. D.; SAMPAIO, A. C. F.; TUDINI, O. G.; ASSUNÇÂO, M. G. T.; ANGELIS NETO, G. de. Evaluation of trees on public roads in the central area of Maringà, state of Paranà:

estimate of waste production and final destination. **Acta. Sci. Agron**, Maringà, v. 29, n. 1, 2007.

ANGIOSPERM PHYLOGENY GROUP (A.P.G) IV. 2016. An update of the Angiosperm Phylogeny Group classification for the orders and families of flowering plants: APG IV.

Botanical Journal of the Linnean Society, London, v. 181, p. 1-20, 2016.

ANTUNES, J. L. F.; CARDOSO, M. R. A. Uses of time series analysis in epidemiological studies. **Epidemiologia e Serviço de Saùde**, 24 (3), 565-576, 2015.

ARACRUZ. Aracruz City Hall. Municipal Environment Department. **Manual of technical recommendations for urban afforestation projects and pruning procedures**. Department of the Environment (SEMAM). Aracruz: [s.n.], 2013.

ARAÙJO, V. C.; BEZERRA, E. S. P.; LIMA NETO, J. A.; ILARIANO, R. N. S.; VASCONCELO, Z. N. F.; VALE, M. B. Study of the use of tree prunings for the production of briquettes in two municipalities in Rio Grande do Norte. In: IX IFRN Scientific Initiation Congress. **Anais ...**Currais Novos/RN, 0673-0679, 2013.

BARATTA JUNIOR, A. P.; MAGALHÂES, L. M. S. Use of tree pruning waste from the city of Rio de Janeiro for composting. **Revista de Ciências Agro- Ambientais**, Alta Floresta, v. 8, n.1, 2010.

BARROS, V. C. C. **Briquettes produced from urban pruning waste and cardboard packaging**. 2013. Monograph (Forestry Engineering). University of Viçosa, Viçosa, 2013.

BATISTEL, L. M.; DIAS, M. A. B.; MARTINS, A. S.; RESENDE, I. S. M. Qualitative and quantitative diagnosis of urban afforestation in the Promissao and Pedro Cardoso neighborhoods, Quirinópolis, Goiás. **Revista da Sociedade Brasileira de Arborizaçâo Urbana**, v.4, n.3, 110-129, 2009.

BENATTI, D. P. B.; TONELLO, K. C.; ADRIANO JUNIOR, F. C.; SILVA, J. M. S.; OLIVEIRA, I. R.; ROLIM, E. N.; FERRAZ, D. L. Urban tree inventory in the municipality of Salto de Pirapora, SP. **Revista Arvore**, v. 36, n. 5, p.887-894, 2012.

BIONDI, D. **Diagnosis of street tree planting in the city of Recife**. 1985. Dissertation (Master's Degree in Forestry Sciences). Federal University of Paranà, Paranà, Curitiba, 1985.

BIONDI, D.; ALTHAUS, M. **Curitiba's street trees**: cultivation and management. Curitiba: FUPEF, 2005. 177p.

BM & FBOVESTA. **Carbon credit auctions**. Sao Paulo. 2012. Available at: <http://www2.bmfbovespa.com.br/Consulta-Leiloes/leiloes-de-credito-de-carbono-login.aspx?idioma=en-br>. Accessed on: April 23, 2017.

BOBROWSKI, R. **Structure and dynamics of street tree planting in Curitiba, Paranà, in the period 1984-2010**. Curitiba, 2010. Dissertation (Master's Degree in Forestry Engineering). Federal University of Paranà, Curitiba, 2010.

BOBROWSKI, R.; BIONDI, D. Characterization of the planting pattern adopted in street tree planting in Curitiba, Paranà. **Revista da Sociedade Brasileira de Arborização Urbana**, v.7, n.3, 2012.

BOX, G. E. P.; JENKINS, G. M. **Time series analysis:** forecasting and control. San Francisco: Holden-Day, 1976.

BRACMORT, K. **Biomass:** Comparison of Definitions in Legislation. Congressional Research Service. 2015. Available at: <https://www.fas.org/sgp/crs/misc/R40529.pdf> Accessed on: April 9, 2017.

BRAND, M. A. **Energy from forest biomass**. Rio de Janeiro. Editora Interciência, 131 p., 2010.

BRAND, M. A.; STAHELIN, T. S. F.; FERREIRA, J. C.; NEVES, M. D. Biomass production for energy generation in *Pinus taeda* L. stands of different ages. **Revista Arvore**, v. 38, n. 2, 2014.

BRAZIL. Law No. 12.305, of August 2, 2010. Establishes the National Solid Waste Policy; amends Law No. 9.605, of February 12, 1998; and makes other provisions. **Diàrio Oficial da Uniao**.3 de agosto. 2010 A. Available at:<http://www.planalto.gov.br/ccivil_03/_ato2007-2010/2010/lei/l12305.htm>. Accessed on: February 1, 2017.

. Law No. 11.445, of January 5, 2007. **Establishes national basic sanitation guidelines and makes other provisions**. Available at: <http://www.planalto.gov.br/ccivil_03/_Ato2007-2010/2007/Lei/L11445.htm#art7>. Accessed on: March 18, 2017.

. MINISTRY OF MINES AND ENERGY. **National Energy Plan 2030:** Biomass. Ministry of Mines and Energy; Collaboration with the Energy Research Company. Brasilia: MME: EPE, v. 12, 2007.

. Decree No. 7.404, of December 23, 2010. Regulates Law No. 12.305, of August 2, 2010, which establishes the National Solid Waste Policy, creates the Interministerial Committee of the National Solid Waste Policy and the Guidance Committee for the Implementation of Reverse Logistics Systems, and makes other provisions. **Official Journal of the Union**. 2010 B. Available at:<http://www.planalto.gov.br/ccivil_03/_ato2007- 2010/2010/decreto/d7404.htm>. Accessed on: February 1, 2017.

BRESSAN, A. A. Decision-making in agricultural futures as time series forecasting models. **Revista de Administraçao de Empresas**, v. 3, n. 1, 2004.

BRUGNARA, G. A. **Forests, wood and housing:** energy and environmental analysis of the production and use of wood as a contribution to the challenge of valuing the Amazon Forest. 2001. Dissertation (Master's Degree in Energy Systems Planning). State University of Campinas, Faculty of Mechanical Engineering, Campinas, 2001.

CABRAL, E. **Considerations on Solid Waste.** Federal University of Cearà, Fortaleza. Class material. Available at:

<http://www.deecc.ufc.br/Download/Gestao_de_Residuos_Solidos_PGTGA/CONSIDERACOES_SOBRE_RESIDUOS_SOLIDOS.pdf>. Accessed on: February 21, 2017.

CAMILO, D. R.; ESPADA, A. L. V.; MARTINS, J. R.F. **Characterization of the management systems for pruning waste and removal of urban trees in the municipalities of the State of Sao Paulo**. Piracicaba. 2008. Supervised internship report. Luiz de Queiroz College of Agriculture, University of Sao Paulo, 2008.

CAMPOS, R. J. **Forecasting time series with applications to electricity consumption series**. 2008. Dissertation (Master's Degree in Electrical Engineering), Federal University of Minas Gerais, Belo Horizonte, f. 110, 2008.

CANADA. Information Center Government of Alberta. **Recommendations for Reducing Leaf and Yard in Aberta**. Alberta Environment and Sustainable Resource Development, 2014. Available at:<http://aep.alberta.ca/waste/documents/ReducingLeafYardWaste-Feb2014.pdf>. Accessed on: February 10, 2017.

CARDOSO-LEITE, E.; FARIA, L. C.; CAPELO, F. F. M.; TONELO, K. C.; CASTELLO, A. C. D.; Floristic composition of urban trees in Sorocaba/SP, Brazil. **Revista da Sociedade Brasileira de Arborizaçao Urbana**, v. 9, n. 1, 2014.

CARVALHO, M.; FREIRE, R. S.; MAGNO, A. H.. **Promotion of sustainability by quantifying and reducing the carbon footprint**: new practices for organizations. In: Global Conference on Global Warming, 2015. Proceedings ... Athens, Greece. proceedings of the global conference on global warming, 2015.

CAVALETT, O. **Analysis of the soybean life cycle**. 2008. Thesis (Doctorate in Food Engineering), State University of Campinas, Faculty of Food Engineering, Campinas - SP. 2008.

CINTRA, T. C. **Energy assessments of native forest species planted in the Médio Paranapanema region, SP**. 2009. 85 f. Dissertation (Master's Degree in Forest Resources). Escola Superior de Agricultura "Luiz de Queiroz", Universidade de Sao Paulo, Piracicaba, 85 f., 2009.

COELHO JUNIOR, L. M.; REZENDE, J. L. P.; CALEGARIO, N.; SILVA, M. L. Longitudinal analysis of charcoal prices in the state of Minas Gerais. **Revista Arvore**, v. 30, n. 3, p. 429-438, 2006.

COELHO JUNIOR, L. M.; REZENDE, J. L. P.; SAFADI, T.; CALEGARIO, N.; Analysis of the temporal behavior of rubber prices in the international market. **Ciência Florestal**, v. 19, n. 3, 2009.

ENERGY COMPANY OF MINAS GERAIS. **Afforestation manual**. Belo Horizonte: CEMIG/Fundaçao Biodiversitas, 2011. . Available at: <http://www.cemig.com.br/sites/imprensa/pt-br/Documents/Manual_Arborizacao_Cemig_Biodiversitas.pdf>. Accessed on: April 10, 2017.

COMPANHIA PARANAENSE DE ENERGIA. **Afforestation of public roads**. Curitiba: PR, 2009. Available at:

<http://www.copel.com/hpcopel/guia_arb/copel_e_a_arborizacao_de_vias_publicas.html>. Accessed on: January 10, 2017.

CORTEZ, C. L. **Study of the potential use of biomass resulting from tree pruning for energy generation: a** case study: AES Eletropaulo. 2011. Thesis (Doctorate in Science). University of Sao Paulo, Sao Paulo, 2011.

COUTO, L. MULLER, M. D.; SILVA JÛNIOR, A. G.; CONDE, L. J. N. Wood pellet production - The case of Bio-Energy in Espirito Santo. **Biomassa & Energia**, v. 1, n. 1, 2004.

CRUZ, S. R. S.; PAULINO, S. R.; Clean development mechanism (CDM) project for landfills and solid waste management in the city of Sao Paulo. In: V Encontro Nacional da Anppas, 2010, Florianópolis. **Anais ...**, V Encontro Nacional da Anppas, 2010. Available at: <http://www.anppas.org.br/encontro5/cd/artigos/GT3-7- 389-20100903200350.pdf>. Accessed on: January 25, 2017.

DANTAS, I. C.; SOUZA, C. M. Arborizaçao urbana na cidade de Campina Grande-PB: Inventàrio e suas espécies. **Revista de Biologia e Ciência da Terra**, vol. 4, n. 2, 2004.

DIAS, J. M. C. S.; SOUZA, D. T.; BRAGA, M.; ONOYAMA, M. M.; MIRANDA, C. H. B.; BARBOSA, P. F. D.; ROCHA, J. D.; **Production of briquettes and pellets from agricultural, agro-industrial and forestry waste**. Embrapa Agroenergia: Document 13, Brasilia, 2012.

DUARTE, A. C. **CDM projects in landfills in Brazil**: an alternative for sustainable development. 125 f. 2006. Dissertation (Master's Degree in Water Resources and Environmental Engineering). Federal University of Paranà. Curitiba, 2006. Available at:

<http://www.ppgerha.ufpr.br/publicacoes/dissertacoes/files/127- Adriana_Carneiro_Duarte.pdf>. Accessed on: Feb. 23, 2017.

ECOINVENT CENTER. **What we do**. Zurich, 2015. Available at: <http://www.ecoinvent.org/about/about.html>. Accessed on: December 16, 2015.

ELOY, E. **Production and quality of biomass from energy forests in northern Rio Grande do Sul**. 2015. 157 f. Thesis (Doctorate in Forest Engineering). Federal University of Paranà, Curitiba, 2015.

EMLUR. SPECIAL MUNICIPAL CLEANING AUTHORITY FOR JOÂO PESSOA. **Information on annual pruning and costs**. Joao Pessoa. 2016.

. **Municipal Integrated Solid Waste Management Plan (PMGIRS)**: Diagnosis and Planning of Urban Cleaning and Solid Waste Management Services. Joao Pessoa. v. 2, 2014 a. Available at:<http://transparencia.joaopessoa.pb.gov.br/dadospublicos/?p=111>. Accessed on: April 13, 2017.

. **Plano Municipal de Gestâo Integrada de Residuos Sólidos:** Prognòstico e Planejamento dos serviços de limpeza urbana e manejo de residuos sólidos. Joao Pessoa, v. 2, 2014 b.

ENERGY RESEARCH COMPANY. **National Energy Balance 2015**. Rio de

January: EPE, 2015. Available at: <https://ben.epe.gov.br/downloads/Relatorio_Final_BEN_2015.pdf>. Accessed on: January 1, 2017.

ENVIRONMENTAL PROTECTION AGENCY. **Food Wast**. 2015. Available at: <http://www3.epa.gov/epawaste/conserve/tools/warm/pdfs/Food_Waste.pdf>. Accessed on: February 10, 2017.

FARIA, R. F.; SOUSA, V. R.; MIRANDA, S. do C. Urban afforestation in the city of Itapuranga, Goiàs. **Revista da Sociedade Brasileira de Arborização Urbana**, v. 9, n. 2, 2014.

FERRAZ. M. C. Inventory of urban trees in the city of Registro-SP. **Revista da Sociedade Brasileira de Arborização Urbana**, v. 7, n. 2, 2012.

FERREIRA, J. V. R. **Product Life Cycle Analysis**. Viseu: Polytechnic Institute of Viseu, 2004.

FIGUEIREDO, J. C. **Estimation of biogas production and energy potential of urban solid waste in Minas Gerais**. 2012, 139 f. Dissertation (Master's Degree in Environmental Systems Analysis and Modeling). Federal University of Minas Gerais, Belo Horizonte, 2012.

FISCHER, S. **Univariant time series**: Box and Jenkins methodology. Porto Alegre: Economics and Statistics Foundation, 1982. 186 p. Available at:

<https://cdn.fee.tche.br/teses/digitalizacao/teses_4.pdf>. Accessed on: April 1, 2017.

FLORIANO, E. P; MULLER, I.; FINGER, C. A. G.; SCHNEIDER, P. R.; Adjustment of traditional models for time series of tree height data. **Ciência Florestal**, v. 16, n. 2, 2006.

FREIRE, R. S; CARVALHO, M.; CARMONA, C. U. M.; MAGNO, A. H.. Perspectives on the implementation of climate change public policies in brazil. In: Global Conference on Global Warming, 2015. **Proceedings...** Athens, Greece. proceedings of the global conference on global warming, 2015.

GALDIANO, G. P. **Life Cycle Inventory of Offset Paper produced in Brazil**. 2006. Dissertation (Master's in Engineering). University of São Paulo, São Paulo, 2006.

GATTI, J. B. **The 4 phases of LCA**. In: MOURAD, A. L.; GARCIA, E. E.C.; VILHENA, A. Life cycle assessment: principles and applications. Campinas: CETEA/CEMPRE. 2002.

GOIÂNIA. Goiânia City Hall. **Goiânia Urban Forestry Master Plan**. Goiânia, 2008. Available at: <http://www.goiania.go.gov.br/download/amma/relatorio_Plano_Diretor.pdf>. Accessed on: January 5, 2017.

GOLVEIA, N. Residuos sólidos urbanos: impactos socioambientais e perspectiva de manejo sustentàvel com inclusão social. **Ciência & Saùde Coletiva**, v. 17, n. 6, p. 1503:1510, 2012.

GREY, G. W. and DENEKE, F. J. **Urban forestry**. New York: John Wiley, 1978.

GUINÉE JB. (ed) **Life Cycle Assessment:** An operation al guide to the ISO Standards**;**

LCA in Perspective; Guide; Operational Annexto Guide. Center for Environmental Science, Leiden University, The Netherlands, 2001.

GUINEA JB. **Handbook on life cycle assessment:** operation al guide to the ISO standards**.** Kluwer Academic Publishers, Boston, 2002.

GUJARATI, D. N; PORTER, D. C. **Basic Economics**. 5th ed. Sao Paulo: BOOKMAN,

HAAREN, R. V.; THEMELIS, N. J.; BARLAZ, M. LCA comparison of windrow composting of yard wastes with use as alternative daily cover (ADC). **Waste Management**, v. 30, f 2649-2656, 2010.

HAUSCHILD, M. Z.; POTTING, J. Spatial differentiation in life cycle impact assessment - the EDP2003 methodology. Guidelines from the Danish environmental protection agency. **Enviromental New**, n°80, Copenhagen, Denmark. 2005.

HUECK, K. **The Forests of South America:** Ecology, Composition and Economic Importance. Sao Paulo: University of Brasilia; Poligono, 465p, 1972.

IEA - INTERNATIONAL ENERGY AGENCY. **Heating Without Global Warming: Market Developments and Policy Considerations for Renewable Heat.**Paris, 2014.

BRAZILIAN Institute of Geography AND STATISTICS. **Municipal population census:** Joao Pessoa. 2016. Available at: <http://cidades.ibge.gov.br/xtras/perfil.php?codmun=250750>. Accessed on: December 31, 2016.

INTERGOVERNMENTAL PANEL ON CLIMATE CHANGE. ^ **Who is the IPCC?** Geneva: Secretariat of the WMO, 2013. Available at: <http://www.ipcc.ch/>. Accessed on: January 1, 2017.

. **History**. Geneva: Secretariat of the WMO, 2015. Available at: <http://www.ipcc.ch/organization/organization_history.shtml>. Accessed on: November 1, 2017.

JARA, E. R. P. The calorific value of some woods found in Brazil. Sao Paulo**: Institute of Technological Research**, 1989.

JOÂO PESSOA. Joao Pessoa City Hall. Joao **Pessoa Sustainable Action Plan 2014.** Joao Pessoa, 2014. Available at: <http://www.joaopessoa.pb.gov.br/>. Accessed on: January 3, 2017.

JOÂO PESSOA. Municipal Department of Communication. **City Hall exceeds target and carries out 9.2 thousand tree prunings in 2012**. Joao Pessoa, 2012. Available at:

<http://www.joaopessoa.pb.gov.br/prefeitura-supera-meta-e-realiza-92-mil-podas-de- arvores-em-2012/>. Accessed on: February 1, 2017.

JOÂO PESSOA. Municipal Planning Department. **Cadastral file of the urban streets of Joao Pessoa/PB.** 2014. Delivered in person.

KLOPFFER, W. **The critical review life cycle assessment studies according to ISO 14040 and 14044:** origin, purpose and practical performance. The International Journal Of Life Cycle Assessment, Heidelberg, p. 1-7. 2012.

LATORRE, M. R. D. O.; CARDOSO, M. R.A. Time series analysis in epidemiology: an introduction to methodological aspects. **Rev. Brasileira de Epidemiologia**, v. 4, n° 3, 2001.

LEAL, L.; BIONDI, D.; ROCHADELLI, R. Cost of implementing and maintaining street tree planting in the city of Curitiba, PR. **Revista Arvore**, v. 32, n. 3, p. 557-565, 2008.

LEVIS, J. W.; BARLAZ, M. A. **Composting Process Model Documentation**; Project Report, North Carolina State University: Raleigh, NC. 2013. Available at: <http://www4.ncsu.edu/~jwlevis/Composting.pdf>Accessed Feb. 11, 2017.

LIMA NETO, E. M.; BIONDI, D.; LEAL, L.; SILVA, F. L. R.; PINHEIRO, F. A. P.

Analysis of the floristic composition of Boa Vista-RR: a subsidy for street tree management.

Revista da Sociedade Brasileira de Arborização Urbana, v. 11, n. 1, p. 58-72.

LINDENMAIER, D. S.; SANTOS, N. O. Arborização Urbana das praças de Cachoeira do Sul-RS-Brasil: Fitogeografia, Diversidade e indice de Areas Verdes. **Botanical Research** No. 59, São Leopoldo: Instituto Anchietano de Pesquisas, 2008.

LINDENMAIER, D. S.; SOUZA, B. S. P.; Road Tree Planting in Cachoeira do Sul/RS: Diversity, Phytogeography and Conflicts with Urban Infrastructure. **Revista da Sociedade Brasileira de Arborização Urbana**, v. 9, n. 1, 2014.

LOBODA, C. R.; ANGELIS, B. L. D. Urban public green areas: concepts, uses and functions. **Ambiência - Revista do Centro de Ciências Agrârias e Ambientais**, Guarapuava/PR, v.1, n.1, p.125-139, 2005.

LOCASTRO, J. K.; RASBOLD, G. G.; PERREIRA, J. S. R.; SOARES, B.; CAXAMBÙ, M. G. Census of urban afforestation in the municipality of Cafeara, Paranà. **Revista da Sociedade Brasileira de Arborização Urbana,** v. 9, n. 3, 2014.

LORENZI, H.; NOBLICK, L. R.; KAHN, F. K.; FERREIRA, E.; **Flora Brasileira Lorenzi: Arecaceae (palm)**. Nova Odessa, SP: Instituto Plantarum, 2010.

LORENZI, H.; SOUZA, H. M.; TORRES, M. A. V.; BACHER, L. B. **Arvores exóticas no Brasil:** Madeiras, ornamentais e aromàtica. Nova Odessa, SP: Instituto Plantarum, 2003.

LUCENA, J. N.; SOUTO, P. C.; CAMANO, J. D. Z.; SOUTO, J. S.; SOUTO, L. S. Afforestation in central flowerbeds in the city of Patos, Paraiba. **Green Journal of Agroecology and Sustainable Development**, v. 10, n. 4, 2015.

MACEDO, L. A.; ROUSSET, P. L. A; VALE, A. T. do. Influence of biomass composition on the condensable yield of vegetable waste torrefaction. **Pesquisa Florestal Brasileira,** v. 34, n.80, 2014.

MAIORANI, E.; WESOLOWSKI, J.B.; MELISINAS, V. A. P. S.; FABRIN, T.C.; GASQUES, L.S. Survey of urban forestry in the municipality of Altônia-PR. **Publicatio UEPG Ciências Biológicas e da Saùde**, Ponta Grossa-PR, v.18, n. 2, p. 101108, 2012.

MARANHO, A. S; PAULA, S. R. P.; LIMA, E.; PAIVA, A. V.; ALVES, A. P.; NASCIMENTO, D. O. Census survey of urban road tree planting in Senador Guimard, Acre. **Revista da Sociedade Brasileira de Arborização Urbana**, v. 7, n. 3, 2012.

MARTINS, C. H. The use of wood from road tree pruning in Maringa/PR. **Revista Verde**, Mossoró, v. 8, n. 2, p. 257 - 267, 2013.

MASTELLA, D. V. **Comparison between the production processes of ceramic and concrete blocks for structural masonry, through life cycle analysis**. 125 f., 2002. Dissertation (Master's

Degree in Civil Engineering). Federal University of Santa Catarina, Florianópolis, 2002.

McPHERSON, E. G.; SIMPSON, G. R. A comparison of municipal forest benefits and costs in Modesto and Santa Monica. **Urban Forestry and Urban Greening** , California, v.1, p.61-74, 2002.

MEIRA, A. M. **Waste management in urban forestry**. 2010. 179 f Thesis (Doctorate in Sciences). Luiz de Queiroz College of Agriculture, University of São Paulo, Piracicaba, 2010.

MELO, E. F. R. Q.; SEVERO, B. M. A. Avenida Brasil (Passo Fundo, Rio Grande do Sul): Vegetation Diversity and Environmental Quality. **Revista da Sociedade Brasileira de Arborização Urbana**, v.5, n.3, 2010.

MEYLER, A.; KENNY, G.; QUINN, T. **Forecasting Irish inflation using ARIMA models**. 1998. 46 p. Available at:

<http://www.centralbank.ie/publications/documents/3RT98.pdf>. Accessed on April 1, 2017.

MILANO, M. S. O planejamento da arborização e as necessidades de manejo e tratamento culturais das árvores de ruas de Curitiba-PR. **Floresta**, v. 17, n. 1E2, 1987.

MIRANDA, Y. C.; MACHADO, M. S.; SAILVA, L. S.; ESTEVAM, R.; MARTINS NETO, F. F.; CAXAMBU, M. G. Qualitative and quantitative analysis of street tree planting in the municipality of Godoy Moreira - PR. **Revista da Sociedade Brasileira de Arborização Urbana,** v. 10, n. 1, 2015.

MORAIS, L. A.; MACHADO, R. R. B. Urban afforestation in the municipality of Timon/MA: inventory, diversity and quali-quantitative diagnosis. **Revista da Sociedade Brasileira de Arborização Urbana,** v. 9, n. 4, 2014.

MOREIRA, J. M. M. A. P. Potential and participation of forests in the energy matrix. **Pesquisa Florestal Brasileira,** v.31, n 68, p. 363-372, 2011.

MORETTIN, P. A.; TOLOI, C. M. C. **Time series analysis**. 2. ed. ver. and ampi. Sao Paulo: Edgar Blücher, 2006.

MORRIS, J.; MATTHEWS, H. S.; MORAWSKI, C. **Review of LCAs on Organics Management Methods & Development of an Environmental Hierarchy**. Information Center Alberta Enviroment. Alberta, 2011. Available at:<http://environment.gov.ab.ca/info/library/8350.pdf>. Accessed on: Feb. 10, 2017.

MOTTER, N.; MÜLLER, N. G. Diagnosis of urban afforestation in the municipality of Tuparendi-RS. **Revista da Sociedade Brasileira de Arborizaçao Urbana,** v.7, n.4, 2012.

NASCIMENTO, M. S.; RODRIGUES, E. R.; SOUZA, C. A.; FARIAS, M. J. B.; PEDERASSI, J.; LIMA, M. S. C. Análisis quali-quantitativa da arborizaçao das áreas públicas do Bairro Centro de Resende, RJ. **Revista da Sociedade Brasileira de Arborizaçao Urbana**, v. 9, n. 4, 2014.

NOWAK, D. J.; WALTON, J. T.; STEVENS, J. C.; CRANE, D. E.; HOEHN, R. E. Effect of plot and sample size on timing and precision of Urban Forest Assessments.

Arboriculture & Urban Forestry, v. 34, n. 06, p. 386-390, 2008.

NUNES, R. L.; MARMONTEL, C. V. F.; RODRIGUES, J. P.; MELO, A. G. C. Qualiquantitative survey of urban afforestation in the Ferraropólis neighborhood in the city of Graça-SP. **Revista da Sociedade Brasileira de Arborizaçao Urbana**, v.8, n.1, 2013.

OLIVEIRA FILHO, P. C.; SILVA, S. V. K. An information system for spatial and decision-making support for urban tree management in the municipality of Guarapuava, Paranà.

Revista da Sociedade Brasileira de Arborizaçao Urbana, v. 5, n. 3, p. 82-96, 2010.

PAIVA, A. V. Aspectos da arborizaçao urbana do Centro de Cosmópolis - SP. **Revista da Sociedade Brasileira de Arborizaçao Urbana**, v. 4, n. 4, p. 17-31, 2009.

PARAIBA. Government of the State of Paraiba. State Secretariat for Water Resources, the Environment and Science and Technology. **Paraiba State Solid Waste Plan - Synthesis Report.** Joao Pessoa, 2014. Available at:

<static.paraiba.pb.gov.br/2013/01/PLANO-ESTADUAL-VERSAO-PRELIMINAR.pdf.>. Accessed on: February 1, 2017.

PARRY, M. M.; SILVA, M. M.; SENA, I. S.; OLIVEIRA, P. M. Composiçao floristica da arborizaçao da cidade de Altamira, Parà. **Revista da Sociedade Brasileira de Arborizaçao Urbana**, v. 7, n. 1, 2012.

PAVAN, M. de C. O.; PARENTE, V. CDM projects in Brazilian landfills: political, socioeconomic and environmental analysis. In: XXX Inter-American Congress of Sanitary and Environmental Engineering, 2006, Punta del Este. Proceedings **...** XXX Inter-American Congress of Sanitary and Environmental Engineering, 2006. Available at: <http://www.cetesb.sp.gov.br/wp-content/uploads/sites/27/2014/01/parente_pavan.pdf.> Accessed on: January 23, 2017.

PERIOTTO, F; MESTRINER, M. M.; HELMANN, A. C.; SANTOS, T. O.; BORTOLOTTI, S. L. Analysis of urban afforestation in the municipality of Medianeira, Paranà.

Revista da Sociedade Brasileira de Arborizaçao Urbana, v. 11, n. 2, 59-74, 2016.

PINDYCK, R. S.; RUBENFIELD, D. L. **Econometric models and economic forecasts**.3nd ed.

New York: McGrawHill, 1991.

PIRES, A.C; RABELO, R. R; XAVIER, J. H. V. **Potential use of Life Cycle Analysis (LCA) associated with organic production concepts applied to family farming**. Cadernos de Ciência e Tecnologia, Brasilia, v. 19, n. 2, p. 149-178, 2002.

PIRES, A.C; RABELO, R. R; XAVIER, J. H. V. **Potential use of Life Cycle Analysis (LCA) associated with organic production concepts applied to family farming**. Cadernos de Ciência e Tecnologia, Brasilia, v. 19, n. 2, p. 149-178, 2002.

PIVETTA, K. F. L.; SILVA FILHO, D. F. **Arborização Urbana**. Boletim Acadêmico: Série Arborização Urbana, Universidade Estadual Paulista, UNESP/FCAP, Jaboticabal, 74 f, 2002.

PRé. **SimaPro Tutorial**. [s.l.]: [s.n.], 2014.

PRETO, E. V.; MORTOZA, G. L. **Electricity generation using biomass**. Monograph (Electrical Engineering). Faculty of Technology, University of Brasilia- UNB, Brasilia, 2010.

QUISSINDO, I. A. B.; OCONOR, E. F.; LUNA, D. P. Evaluation of tree vegetation in the main streets of the city of Huambo-Angola. **Revista da Sociedade Brasileira de Arborizaçao Urbana**, v. 11, n. 1, p. 43-57, 2016.

RABER, A.P.; REBELATO, G.S. Arborização viària do municipio de Colorado, RS - BRASIL: análisis quali-quantitativa. **Revista da Sociedade Brasileira de Arborização Urbana**, v.5, n.1, 2010.

RECIFE. Secretariat for the Environment and Sustainability (SEMAS). **Afforestation manual:** guidance and technical procedures for implementation in the city of Recife. Secretariat for the Environment and Sustainability (SEMAS), Recife: [s.n.], 2013.

REFLORA. **List of Species of the Flora of Brazil.** Rio de Janeiro Botanical Garden Research Institute. 2017. Access

at:<http://reflora.jbrj.gov.br/jabot/listaBrasil/ConsultaPublicaUC/ConsultaPublicaUC.do> . Accessed on: February 10, 2017.

REICHERT, G. A.; MENDES, C. A. B. Life cycle assessment and decision support in integrated management and sustainability of urban solid waste. **Eng. Sanit.**

Ambiental, v. 19, n. 13, 2014.

RIZZI, C. A. A **questão da participação da comunidade do Distrito de Perus (São Paulo/Brazil), no projeto MDL Aterro Bandeirantes, Confins** [Enligne], 2011.

Available at:

<https://confins.revues.org/6870?lang=pthttp://www.revistas.sp.senac.br/index.php/ITF/article/viewFile/97/122>. Accessed on: January 25, 2017.

ROCHA, A. T.; SANTOS, P. S.; NETO, S. N. O. Arborização de vias publica em Nova Iguaçu, RJ: o caso dos Bairros Rancho Novo e Centro. **Revista Arvore**, v. 28, n. 4, p. 599607, 2004.

ROSARIO, L. M. **Briquetting for the use of sawmill waste**. 2011. Monograph (Degree in Industrial Wood Engineering). Federal University of Espirito Santo, Jerônimo Monteiro, 2012.

ROSSETTI, A. I. N.; PELLEGRINO, P. R. M; TAVARES, A. R. Trees and their interfaces in the urban environment. **Revista da Sociedade Brasileira de Arborização Urbana,** Piracicaba - SP, v.5, n.1, p.1-24, 2010.

ROVERE, E. L. La; COSTA, C. VALLE; DUBEUX, C. B. S. **Landfill in Brazil and the Clean Development Mechanism (CDM)**: Opportunity to promote socio-environmental development. Web-Resol, Rio Bonito, 2006. Available at: <http://www.web-resol.org/textos/28-La%20Rovere%20E.pdf.> Accessed on: October 22, 2016.

RUMÂO, A. S. **Generation of Power and Electric Energy from the Gasification of Biomass Waste.** Federal University of Paraiba. Thesis (Doctorate in Mechanical Engineering), Postgraduate Program in Mechanical Engineering, Federal University of Paraiba, 130 p., 2013.

SALVI, L.T.; HARDT, L.P.A.; ROVEDDER, C.E.; FONTANA, C.S. Tree planting along streets - Túneis Verdes- in Porto Alegre, RS, Brazil: quantitative and qualitative evaluation. **Revista Arvore**, v.35, n.2, p.233-243. 2011.

SANCHOTENE, M. C.C. Desenvolvimento e perspectivas da arborização urbana no Brasil. In: Congresso Brasileiro de Arborização Urbana, 2, 1994. Sâo Luis - MA. **Proceedings...** Sâo Luis, Sociedade Brasileira de Arborizaçâo Urbana; 1994.

SANTAMOUR JUNIOR, F.S. Trees for urban planting: diversity uniformity, and common sense. **Proceedings...** In: METRIACONFERENCE, 7., 1990, Lisle. Proceedings. Lisle: p.57-66. 1990.

SANTOS JUNIOR, A.; COSTA, L. M.; Species used in the urban afforestation of Bairro Santiago, Ji-Paranà/RO. **Revista da Sociedade Brasileira de Arborizaçâo Urbana**, v. 9, n. 1, 2014.

SANTOS, A. R.; BERGALLO, H. G.; ROCHA, C. F. D. Alien urban landscape. **Revista Ciência Hoje**, v. 41, n. 245, p. 68-70, 2008.

SANTOS, J. R. S.. **Study of torrefied biomass from eucalyptus forest residues and sugar cane bagasse for energy purposes**. 2012. Dissertation (Master of Science, Program: Forest Resources). Universidade de Sâo Paulo, Escola Superior de Agricultura "Luiz de Queiroz", Piracicaba, p. 86, 2012.

SANTOS, J. S.; SANTOS, G.D. Microclimatic Study in Representative Points of the Urban Network of the City of Joâo Pessoa\PB: An Evaluation of the Thermal Field. **Revista Brasileira de Geografia Fisica**, v. 6, n. 5, p. 1430-1448, 2013.

SANTOS, N. R. Z; TEIXEIRA, I. F. **Arborização de Vias Pùblicas:** Environment X Vegetation, Porto Alegre: Editora Pallotti, 2001.

SANTOS, R. C.; CARNEIRO, A. C. O.; CASTRO, R. V. O.; PIMENTA, A. S.; CASTRO,

A. F. N. M.; MARINHO, I. V.; VILLAS BOAS, M. A.; Briquetting potential of forest residues from the Seridó region of Rio Grande do Norte. **Pesquisa Florestal Brasileira**, v. 31, n. 68, 2011.

SANTOS, T. O. B.; LISBOA, C. M. C. A.; CARVALHO, F. G. Anàlise da arborização viària do bairro de Petrópolis, Natal, RN: Uma abordagem para diagnóstico e planejamento da flora urbana. **Revista da Sociedade Brasileira de Arborização Urbana**, v. 7, n. 4, 2012.

SÂO PAULO (Capital). Municipal Department of Greenery and the Environment. **Technical Manual for Urban Afforestation**. Sâo Paulo, 2015. Available at:

<http://www.prefeitura.sp.gov.br/cidade/secretarias/meio_ambiente/publicacoes_svma/ind ex.php?p=188452>. Accessed on: January 07, 2017.

SARTORI, R. A.; BALDERI, A. P. Inventory of urban afforestation in the municipality of Socorro-SP and proposal for an index of damage to city infrastructure. **Revista da Sociedade Brasileira de Arborização Urbana**, v.6, n.4, 2011.

SAWIN, J. L.; BHATTACHARYA, S. C.; GALÀN, E. M.; McCRONE, A.; MOOMAW, W. R.; SONNTAG-O'BRIEN, V.; SVERRISSON, F.; CHAWLA, K.; MUSOLINO, E.;

SKEEN, J.; MARTINOT, E. **Renewables 2012 Global Status Report**. Paris: REN 21, 2012. Available at:<http://www.ren21.net/about-ren21/about-us/>. Accessed on: January 13, 2017.

SCHALLENBERGER, L. S.; ARAÙJO, A. J; ARAÙJO, M. N.; DAINER. L. J.;

MACHADO, G. O. Evaluation of the condition of urban trees in the main parks and squares in the municipality of Irati/PR. **Revista da Sociedade Brasileira de Arborização Urbana**, v.5, n.2, 105-123, 2010.

SILVA FILHO, D. F.; PIZETTA, P. U. C.; ALMEIDA, j. B. S. A.; PIVETTA, K. F. L.; FERRAOUDO, A. S. Related database for registration, evaluation and management of trees on public roads. **Revista Arvore**, v. 26, n. 5, p. 629-642, 2002.

SILVA, E. **Cultivated biomass for energy production:** comparative study between elephant grass and eucalyptus with the incorporation of solar energy in drying. 2012.

Dissertation (Master of Science in Energy Engineering). Federal University of Itajubà, Itajubà, p. 85, 2012 a.

SILVA, D. A. L. **Life cycle assessment of MDF wood panel production in Brazil**. 2012. Dissertaçâo (Master's Degree in Materials Science and Engineering), Universidade de Sâo Paulo, Sâo Carlos, p. 207, 2012 b.

SILVA, T. G.; LEITA, E. C.; TONELLO, K. C. Inventory of urban afforestation in the municipality of Araçoiaba da Serra, SP. **Revista da Sociedade Brasileira de Arborização Urbana**, v. 9, n. 4, 2014.

SIMAPRO. **SimaPro Data base 8**: Manual Methods Library. California, 2016. Available at: <https://www.pre-sustainability.com/simapro-database-and-methods-library>. Accessed on: January 4, 2017.

SNSA - NATIONAL SECRETARIAT FOR ENVIRONMENTAL SANITATION. System

National Sanitation Information Database. **Diagnosis of urban solid waste management - 2014.** Part 2 - Table of Information and Indicators. Brasilia: MCIDADES. SNSA, 2016.

SOARES, C. P. B.; NETO FRANCISCO, P.; SOUZA, A. L. **Dendrometry and** forest **inventory**. 2. ed. Viçosa: Universidade Federal de Viçosa, Viçosa, MG. 2011.

SOARES, N. S.; SILVA, M. L.; LIMA, J. E.; CORDEIRO, S. A. Forecast analysis of the price of natural rubber in Brazil. **Scientia Forestalis**, v. 36, n. 80, p. 285-294, 2008.

SOARES, N. S.; SILVA, M. L.; REZENDE, J. L. P.; LIMA, J. E.; CARVALHO, K. H. A.; Elaboration of a model for predicting the price of *Eucalyptus spp.* wood **Cerne**, v. 16, n. 1, p. 41-52, 2010.

SOUZA, A.M.; NACHTERGAELE, M.F. and CARBONI, M. Inventàrio Florestal da arborizaçao do municipio de Jaù/SP. Jaù: Instituto Pró-Terra & Secretaria do Meio Ambiente - SEMEIA. **Technical report**, 2004.

SOUZA, G. D.; RIBEIRO, W. C. Nova Gerar: Brazil's pioneering CDM experience. **Cronos**, v. 10, n. 2, p. 15-34, 2009.

SOUZA, P. F. de; BOURSCHEID, C. B.; POMPÉO, P. N.; STANG, M. B.; MANFROI, J.; RODRIGUES, M. D. S.; SILVA, A. C. da; HIGUCHI, P. Inventory and Recommendations for the Arborization of the City Center of Sao Joaquim, SC. **Revista da Sociedade Brasileira de Arborizaçao Urbana**, v. 9, n. 4, 2014.

SOUZA, J. F.; SILVA, R. M.; SILVA, A. M. Influence of land use and occupation on surface temperature: the case study of Joao Pessoa-PB. **Ambiente Construido**, v. 16, n. 1, p. 21-37, 2016.

TÔRRES FILHO, A. **Technical and environmental feasibility of using wood waste to produce an alternative fuel**. 2007, 61 f. Dissertation (Master's Degree in Sanitation, Environment and Water Resources). Federal University of Minas Gerais, Belo Horizonte, 2005.

TOSCAN, M. A. G.; RICKLI, H. C.; BARTINICK, D. SANTOS, D. S.; ROSSA, D. Inventory and analysis of the afforestation of the Vila Yolanda neighborhood, in the municipality of Foz do Iguaçu-PR. **Revista da Sociedade Brasileira de Arborizaçao Urbana**, v. 5, n. 3, p. 165184, 2010.

TROPICOS. **Missouri Botanical Garden**. 2016. Available at: <http://tropicos.org>. Accessed on: 01/10/2016.

VALE, N. F. L.; SOUSA, G. S.; MATA, M. F.; BRAGA, P. E. T. Inventàrio Arbòreo do Parque da Cidade do Municipio de Sobral, Cearà. **Revista da Sociedade Brasileira de Arborizaçao Urbana**, v.6, n.4, 2011.

VALINHOS. Valinhos City Hall. **Integrated Solid Waste Management Plan for the Municipality of Valinhos-SP**. Valinhos, 2011. Available at: <http://www.valinhos.sp.gov.br/portal/arquivos/planejamento/PGIRS_- _Verso_Preliminar_2.pdf>. Accessed on: February 10, 2017.

VENABLE, W. N.; RIPLEY, B. D. **Modern applied statistic with S-PLUS**. 3. ed. New York: Springer-Verlag, 1999. 501 p.

VIEIRA, G. E. G.; NUNES, A. P.; TEIXEIRA, L. F.; COLEN, A. G. N. Biomass: a view of pyrolysis processes. **Revista Liberato**, Nova Hamburgo, v. 15, n. 24, 2014.

ZHANG, S. **What is the Hisghest and Best Use of Organic Solid Waste**: Production of Compostor Production of Energy? Greenest City Scholar, 2012. Available at: <https://sustain.ubc.ca/sites/sustain.ubc.ca/files/Zero%20Waste%20%20-%20Siduo%20Zhang%20-%20Compost%20vs%20%20Energy%20Production.pdf>. Accessed on: February 10, 2017.

Printed by Books on Demand GmbH, Norderstedt / Germany